Scientific Computation

T0155795

Springer
Berlin
Heidelberg
New York
Hong Kong
London
Milan
Paris
Tokyo

Physics and Astronomy ONLINE LIBRARY

http://www.springer.de/phys/

H. Lomax   T. H. Pulliam   D. W. Zingg

Fundamentals of Computational Fluid Dynamics

With 25 Figures

Springer

H. Lomax (1922–1999)

T.H. Pulliam
NASA Ames Research Center
Moffett Field, CA 94035, USA

Professor D.W. Zingg
University of Toronto
Institute for Aerospace Studies
4925 Dufferin st.
Toronto, Ontario M3H 5T6, Canada

Library of Congress Cataloging-in-Publication Data.
Lomax, Harvard. Fundamentals of computational fluid dynamics / H. Lomax. T. H. Pulliam, D. W. Zingg. p.cm. – (Scientific computation, ISSN 1434-8322) Includes bibliographic references and index. ISBN 3-540-41607-2 (acid-free paper) 1. Fluid dynamics–Data processing. I. Pulliam, T. H. (Thomas H.), 1951– . II. Zingg, D. W. (David W.), 1958– . III. title. IV. Series.   QA911.L66 2001   532'05-dc21   2001020294

First Edition 2001. Corrected Second Printing 2003.
ISSN 1434-8322
ISBN 978-3-642-07484-4

Springer-Verlag Berlin Heidelberg New York
a member of BertelsmannSpringer Science+Business Media GmbH

http://www.springer.de

© Springer-Verlag Berlin Heidelberg 2001
Softcover reprint of the hardcover 1st edition 2001

Data conversion by LE-TeX, Leipzig
Cover design: *design & production* GmbH, Heidelberg
Printed on acid-free paper   55/3141/RM - 5 4 3 2 1 0

To our families:
Joan, Harvard Laird, James, and Melinda
Carol, Teela, Troy, Merle Lynne, and Richie
Catherine, Natasha, Genevieve, and Trevor

Preface

The field of computational fluid dynamics (CFD) has already had a significant impact on the science and engineering of fluid dynamics, ranging from a role in aircraft design to enhancing our understanding of turbulent flows. It is thus not surprising that there exist several excellent books on the subject. We do not attempt to duplicate material which is thoroughly covered in these books. In particular, our book does not describe the most recent developments in algorithms, nor does it give any instruction with respect to programming. Neither turbulence modelling nor grid generation are covered. This book is intended for a reader who seeks a deep understanding of the fundamental principles which provide the foundation for the algorithms used in CFD. As a result of this focus, the book is suitable for a first course in CFD, presumably at the graduate level.

The underlying philosophy is that the theory of linear algebra and the attendant eigenanalysis of linear systems provide a mathematical framework to describe and unify most numerical methods in common use for solving the partial differential equations governing the physics of fluid flow. This approach originated with the first author during his long and distinguished career as Chief of the CFD Branch at the NASA Ames Research Center. He believed that a thorough understanding of sophisticated numerical algorithms should be as natural to a physicist as a thorough understanding of calculus. The material evolved from the notes for the course *Introduction to Computational Fluid Dynamics* taught by the first author at Stanford University from 1970 to 1994. The second author has taught the course since 1994, while the third author has taught a similar course at the University of Toronto since 1988.

The material is generally suitable for a typical graduate course of roughly 15 weeks. It is unlikely that the entire book can be covered in the lectures, and it is left to the instructor to set priorities. One approach which has worked well is to devote one lecture to each chapter, covering as much material as possible and leaving the remainder as reading material. The most essential material is in Chapters 2–4 and 6–8. Most of the chapters should be covered in the sequence used in the book. Exceptions include Chapter 5, which can be treated later in the course if desired, as well as Chapters 11–13, which can be covered anytime after Chapter 8. The mathematics background associated with a typical undergraduate degree in engineering should be sufficient to understand the material.

The second and third authors would like to acknowledge their enormous debt to Harvard Lomax, both as a mentor and a friend. Sadly, Harvard passed away in May 1999. We present this book as further evidence of his outstanding legacy in CFD.

Moffett Field and Toronto,   *Thomas H. Pulliam*
March 2001   *David W. Zingg*

Contents

1. Introduction

1.1 Motivation

The material in this book originated from attempts to understand and sys-
temize numerical solution techniques for the partial differential equations
governing the physics of fluid flow. As time went on and these attempts began to crystallize, underlying constraints on the nature of the material began
to form. The principal such constraint was the demand for unification. Was
there one mathematical structure which could be used to describe the behavior and results of most numerical methods in common use in the field of
fluid dynamics? Perhaps the answer is arguable, but the authors believe the
answer is affirmative and present this book as justification for that belief.
The mathematical structure is the theory of linear algebra and the attendant
eigenanalysis of linear systems.

The ultimate goal of the field of computational fluid dynamics (CFD) is
to understand the physical events that occur in the flow of fluids around and
within designated objects. These events are related to the action and inter-
action of phenomena such as dissipation, diffusion, convection, shock waves,
slip surfaces, boundary layers, and turbulence. In the field of aerodynamics,
all of these phenomena are governed by the compressible Navier–Stokes equations. Many of the most important aspects of these relations are nonlinear
and, as a consequence, often have no analytic solution. This, of course, motivates the numerical solution of the associated partial differential equations.
At the same time it would seem to invalidate the use of linear algebra for the
classification of the numerical methods. Experience has shown that such is
not the case.

As we shall see in a later chapter, the use of numerical methods to solve
partial differential equations introduces an approximation that, in effect, can
change the form of the basic partial differential equations themselves. The
new equations, which are the ones actually being solved by the numerical
process, are often referred to as the modified partial differential equations.
Since they are not precisely the same as the original equations, they can, and
probably will, simulate the physical phenomena listed above in ways that
are not exactly the same as an exact solution to the basic partial differential
equation. Mathematically, these differences are usually referred to as truncation errors. However, the theory associated with the numerical analysis of

fluid mechanics was developed predominantly by scientists deeply interested in the physics of fluid flow and, as a consequence, these errors are often identified with a particular physical phenomenon on which they have a strong effect. Thus methods are said to have a lot of "artificial viscosity" or said to be highly dispersive. This means that the errors caused by the numerical approximation result in a modified partial differential equation having additional terms that can be identified with the physics of dissipation in the first case and dispersion in the second. There is nothing wrong, of course, with identifying an error with a physical process, nor with deliberately directing an error to a specific physical process, as long as the error remains in some engineering sense "small". It is safe to say, for example, that most numerical methods in practical use for solving the nondissipative Euler equations create a modified partial differential equation that produces some form of dissipation. However, if used and interpreted properly, these methods give very useful information.

Regardless of what the numerical errors are called, if their effects are not thoroughly understood and controlled, they can lead to serious difficulties, producing answers that represent little, if any, physical reality. This motivates studying the concepts of stability, convergence, and consistency. On the other hand, even if the errors are kept small enough that they can be neglected (for engineering purposes), the resulting simulation can still be of little practical use if inefficient or inappropriate algorithms are used. This motivates studying the concepts of stiffness, factorization, and algorithm development in general. All of these concepts we hope to clarify in this book.

1.2 Background

The field of computational fluid dynamics has a broad range of applicability. Independent of the specific application under study, the following sequence of steps generally must be followed in order to obtain a satisfactory solution.

1.2.1 Problem Specification and Geometry Preparation

The first step involves the specification of the problem, including the geometry, flow conditions, and the requirements of the simulation. The geometry may result from measurements of an existing configuration or may be associated with a design study. Alternatively, in a design context, no geometry needs to be supplied. Instead, a set of objectives and constraints must be specified. Flow conditions might include, for example, the Reynolds number and Mach number for the flow over an airfoil. The requirements of the simulation include issues such as the level of accuracy needed, the turnaround time required, and the solution parameters of interest. The first two of these requirements are often in conflict and compromise is necessary. As an example of solution parameters of interest in computing the flowfield about an airfoil, one may be interested in (i) the lift and pitching moment only, (ii) the drag as well as the lift and pitching moment, or (iii) the details of the flow at some specific location.

1.2.2 Selection of Governing Equations and Boundary Conditions

Once the problem has been specified, an appropriate set of governing equations and boundary conditions must be selected. It is generally accepted that the phenomena of importance to the field of continuum fluid dynamics are governed by the conservation of mass, momentum, and energy. The partial differential equations resulting from these conservation laws are referred to as the Navier–Stokes equations. However, in the interest of efficiency, it is always prudent to consider solving simplified forms of the Navier–Stokes equations when the simplifications retain the physics which are essential to the goals of the simulation. Possible simplified governing equations include the potential-flow equations, the Euler equations, and the thin-layer Navier–Stokes equations. These may be steady or unsteady and compressible or incompressible. Boundary types which may be encountered include solid walls, inflow and outflow boundaries, periodic boundaries, symmetry boundaries, etc. The boundary conditions which must be specified depend upon the governing equations. For example, at a solid wall, the Euler equations require flow tangency to be enforced, while the Navier–Stokes equations require the no-slip condition. If necessary, physical models must be chosen for processes which cannot be simulated within the specified constraints. Turbulence is an example of a physical process which is rarely simulated in a practical context (at the time of writing) and thus is often modeled. The success of a simulation depends greatly on the engineering insight involved in selecting the governing equations and physical models based on the problem specification.

1.2.3 Selection of Gridding Strategy and Numerical Method

Next a numerical method and a strategy for dividing the flow domain into cells, or elements, must be selected. We concern ourselves here only with numerical methods requiring such a tessellation of the domain, which is known as a grid, or mesh. Many different gridding strategies exist, including structured, unstructured, hybrid, composite, and overlapping grids. Furthermore, the grid can be altered based on the solution in an approach known as solution-adaptive gridding. The numerical methods generally used in CFD can be classified as finite-difference, finite-volume, finite-element, or spectral methods. The choices of a numerical method and a gridding strategy are strongly interdependent. For example, the use of finite-difference methods is typically restricted to structured grids. Here again, the success of a simulation can depend on appropriate choices for the problem or class of problems of interest.

1.2.4 Assessment and Interpretation of Results

Finally, the results of the simulation must be assessed and interpreted. This step can require post-processing of the data, for example calculation of forces and moments, and can be aided by sophisticated flow visualization tools and error estimation techniques. It is critical that the magnitude of both numerical and physical-model errors be well understood.

1.3 Overview

It should be clear that successful simulation of fluid flows can involve a wide range of issues from grid generation to turbulence modelling to the applicability of various simplified forms of the Navier–Stokes equations. Many of these issues are not addressed in this book. Instead we focus on numerical methods, with emphasis on finite-difference and finite-volume methods for the Euler and Navier–Stokes equations. Rather than presenting the details of the most advanced methods, which are still evolving, we present a foundation for developing, analyzing, and understanding such methods.

Fortunately, to develop, analyze, and understand most numerical methods used to find solutions for the complete compressible Navier–Stokes equations, we can make use of much simpler expressions, the so-called "model" equations. These model equations isolate certain aspects of the physics contained in the complete set of equations. Hence their numerical solution can illustrate the properties of a given numerical method when applied to a more complicated system of equations which governs similar physical phenomena. Although the model equations are extremely simple and easy to solve, they have been carefully selected to be representative, when used intelligently, of difficulties and complexities that arise in realistic two- and three-dimensional fluid flow simulations. We believe that a thorough understanding of what happens when numerical approximations are applied to the model equations is a major first step in making confident and competent use of numerical approximations to the Euler and Navier–Stokes equations. As a word of caution, however, it should be noted that, although we can learn a great deal by studying numerical methods as applied to the model equations and can use that information in the design and application of numerical methods to practical problems, there are many aspects of practical problems which can only be understood in the context of the complete physical systems.

1.4 Notation

The notation is generally explained as it is introduced. Bold type is reserved for real physical vectors, such as velocity. The vector symbol ⁻ is used for the vectors (or column matrices) which contain the values of the dependent variable at the nodes of a grid. Otherwise, the use of a vector consisting of a collection of scalars should be apparent from the context and is not identified by any special notation. For example, the variable u can denote a scalar Cartesian velocity component in the Euler and Navier–Stokes equations, a scalar quantity in the linear convection and diffusion equations, and a vector consisting of a collection of scalars in our presentation of hyperbolic systems. Some of the abbreviations used throughout the text are listed and defined below.

PDE	Partial differential equation
ODE	Ordinary differential equation
OΔE	Ordinary difference equation
RHS	Right-hand side
P.S.	Particular solution of an ODE or system of ODE's
S.S.	Fixed (time-invariant) steady-state solution
kD	k-dimensional space
$\left(\vec{bc}\right)$	Boundary conditions, usually a vector
$O(\alpha)$	A term of order (i.e. proportional to) α
$\Re(z)$	The real part of a complex number z

2. Conservation Laws
and the Model Equations

We start out by casting our equations in the most general form, the integral conservation-law form, which is useful in understanding the concepts involved in finite-volume schemes. The equations are then recast into divergence form, which is natural for finite-difference schemes. The Euler and Navier–Stokes equations are briefly discussed in this chapter. The main focus, though, will be on representative model equations, in particular, the convection and diffusion equations. These equations contain many of the salient mathematical and physical features of the full Navier–Stokes equations. The concepts of convection and diffusion are prevalent in our development of numerical methods for computational fluid dynamics, and the recurring use of these model equations allows us to develop a consistent framework of analysis for consistency, accuracy, stability, and convergence. The model equations we study have two properties in common. They are linear partial differential equations (PDE's) with coefficients that are constant in both space and time, and they represent phenomena of importance to the analysis of certain aspects of fluid dynamics problems.

2.1 Conservation Laws

Conservation laws, such as the Euler and Navier–Stokes equations and our model equations, can be written in the following integral form:

$$\int_{V(t_2)} Q \mathrm{d}V - \int_{V(t_1)} Q \mathrm{d}V + \int_{t_1}^{t_2} \oint_{S(t)} \mathbf{n} \cdot \mathbf{F} \mathrm{d}S \mathrm{d}t = \int_{t_1}^{t_2} \int_{V(t)} P \mathrm{d}V \mathrm{d}t . \quad (2.1)$$

In this equation, Q is a vector containing the set of variables which are conserved, e.g. mass, momentum, and energy, per unit volume. The equation is a statement of the conservation of these quantities in a finite region of space with volume $V(t)$ and surface area $S(t)$ over a finite interval of time $t_2 - t_1$. In two dimensions, the region of space, or cell, is an area $A(t)$ bounded by a closed contour $C(t)$. The vector $\mathbf{n}$ is a unit vector normal to the surface pointing outward, $\mathbf{F}$ is a set of vectors, or tensor, containing the flux of Q per unit area per unit time, and P is the rate of production of Q per unit volume per unit time. If all variables are continuous in time, then (2.1) can be rewritten as

$$\frac{\mathrm{d}}{\mathrm{d}t} \int_{V(t)} Q \mathrm{d}V + \oint_{S(t)} \mathbf{n} \cdot \mathbf{F} \mathrm{d}S = \int_{V(t)} P \mathrm{d}V . \qquad (2.2)$$

Those methods which make various numerical approximations of the integrals in (2.1) and (2.2) and find a solution for Q on that basis are referred to as *finite-volume methods*. Many of the advanced codes written for CFD applications are based on the finite-volume concept.

On the other hand, a partial derivative form of a conservation law can also be derived. The divergence form of (2.2) is obtained by applying Gauss's theorem to the flux integral, leading to

$$\frac{\partial Q}{\partial t} + \nabla \cdot \mathbf{F} = P , \qquad (2.3)$$

where $\nabla \cdot$ is the well-known divergence operator given, in Cartesian coordinates, by

$$\nabla \cdot \equiv \left(\mathbf{i} \frac{\partial}{\partial x} + \mathbf{j} \frac{\partial}{\partial y} + \mathbf{k} \frac{\partial}{\partial z} \right) \cdot , \qquad (2.4)$$

and $\mathbf{i}, \mathbf{j}$, and $\mathbf{k}$ are unit vectors in the x, y, and z coordinate directions, respectively. Those methods which make various approximations of the derivatives in (2.3) and find a solution for Q on that basis are referred to as *finite-difference methods*.

2.2 The Navier–Stokes and Euler Equations

The Navier–Stokes equations form a coupled system of nonlinear PDE's describing the conservation of mass, momentum, and energy for a fluid. For a Newtonian fluid in one dimension, they can be written as

$$\frac{\partial Q}{\partial t} + \frac{\partial E}{\partial x} = 0 \qquad (2.5)$$

with

$$Q = \begin{bmatrix} \rho \\ \rho u \\ e \end{bmatrix} , \quad E = \begin{bmatrix} \rho u \\ \rho u^2 + p \\ u(e + p) \end{bmatrix} - \begin{bmatrix} 0 \\ \frac{4}{3} \mu \frac{\partial u}{\partial x} \\ \frac{4}{3} \mu u \frac{\partial u}{\partial x} + \kappa \frac{\partial T}{\partial x} \end{bmatrix} , \qquad (2.6)$$

where ρ is the fluid density, u is the velocity, e is the total energy per unit volume, p is the pressure, T is the temperature, μ is the coefficient of viscosity, and κ is the thermal conductivity. The total energy e includes internal energy per unit volume $\rho \epsilon$ (where ϵ is the internal energy per unit mass) and kinetic energy per unit volume $\rho u^2/2$. These equations must be supplemented by relations between μ and κ and the fluid state as well as an equation of state, such as the ideal gas law. Details can be found in many textbooks. Note that the convective fluxes lead to first derivatives in space, while the viscous and heat conduction terms involve second derivatives. This form of the equations is called *conservation-law* or *conservative* form. Non-conservative forms can be obtained by expanding derivatives of products using the product rule or by introducing different dependent variables, such as u and p. Although non-conservative forms of the equations are *analytically* the same as the above form, they can lead to quite different *numerical* solutions in terms of shock strength and shock speed, for example. Thus the conservative form is appropriate for solving flows with features such as shock waves.

Many flows of engineering interest are steady (time-invariant), or at least may be treated as such. For such flows, we are often interested in the steady-state solution of the Navier–Stokes equations, with no interest in the transient portion of the solution. The steady solution to the one-dimensional Navier–Stokes equations must satisfy

$$\frac{\partial E}{\partial x} = 0 . \tag{2.7}$$

If we neglect viscosity and heat conduction, the Euler equations are obtained. In two-dimensional Cartesian coordinates, these can be written as

$$\frac{\partial Q}{\partial t} + \frac{\partial E}{\partial x} + \frac{\partial F}{\partial y} = 0 \tag{2.8}$$

with

$$Q = \begin{bmatrix} q_1 \\ q_2 \\ q_3 \\ q_4 \end{bmatrix} = \begin{bmatrix} \rho \\ \rho u \\ \rho v \\ e \end{bmatrix} , \quad E = \begin{bmatrix} \rho u \\ \rho u^2 + p \\ \rho u v \\ u(e+p) \end{bmatrix} , \quad F = \begin{bmatrix} \rho v \\ \rho u v \\ \rho v^2 + p \\ v(e+p) \end{bmatrix} , \tag{2.9}$$

where u and v are the Cartesian velocity components. Later on we will make use of the following form of the Euler equations as well:

$$\frac{\partial Q}{\partial t} + A\frac{\partial Q}{\partial x} + B\frac{\partial Q}{\partial y} = 0 . \tag{2.10}$$

The matrices $A = \frac{\partial E}{\partial Q}$ and $B = \frac{\partial F}{\partial Q}$ are known as the flux Jacobians. The flux vectors given above are written in terms of the *primitive* variables, ρ, u, v, and p. In order to derive the flux Jacobian matrices, we must first write the flux vectors E and F in terms of the *conservative* variables, q_1, q_2, q_3, and q_4, as follows:

$$E = \begin{bmatrix} E_1 \\ E_2 \\ E_3 \\ E_4 \end{bmatrix} = \begin{bmatrix} q_2 \\ (\gamma-1)q_4 + \dfrac{3-\gamma}{2}\dfrac{q_2^2}{q_1} - \dfrac{\gamma-1}{2}\dfrac{q_3^2}{q_1} \\ \dfrac{q_3 q_2}{q_1} \\ \gamma\dfrac{q_4 q_2}{q_1} - \dfrac{\gamma-1}{2}\left(\dfrac{q_2^3}{q_1^2} + \dfrac{q_3^2 q_2}{q_1^2}\right) \end{bmatrix} \tag{2.11}$$

$$F = \begin{bmatrix} F_1 \\ F_2 \\ F_3 \\ F_4 \end{bmatrix} = \begin{bmatrix} q_3 \\[2mm] \dfrac{q_3 q_2}{q_1} \\[3mm] (\gamma - 1)q_4 + \dfrac{3 - \gamma}{2}\dfrac{q_3^2}{q_1} - \dfrac{\gamma - 1}{2}\dfrac{q_2^2}{q_1} \\[3mm] \gamma \dfrac{q_4 q_3}{q_1} - \dfrac{\gamma - 1}{2}\left(\dfrac{q_2^2 q_3}{q_1^2} + \dfrac{q_3^3}{q_1^2}\right) \end{bmatrix}. \qquad (2.12)$$

We have assumed that the pressure satisfies $p = (\gamma - 1)[e - \rho(u^2 + v^2)/2]$ from the ideal gas law, where γ is the ratio of specific heats, c_p/c_v. From this it follows that the flux Jacobian of E can be written in terms of the conservative variables as

$$A = \frac{\partial E_i}{\partial q_j} = \begin{bmatrix} 0 & 1 & 0 & 0 \\[2mm] a_{21} & (3-\gamma)\left(\dfrac{q_2}{q_1}\right) & (1-\gamma)\left(\dfrac{q_3}{q_1}\right) & \gamma - 1 \\[3mm] -\left(\dfrac{q_2}{q_1}\right)\left(\dfrac{q_3}{q_1}\right) & \left(\dfrac{q_3}{q_1}\right) & \left(\dfrac{q_2}{q_1}\right) & 0 \\[3mm] a_{41} & a_{42} & a_{43} & \gamma\left(\dfrac{q_2}{q_1}\right) \end{bmatrix}, \quad (2.13)$$

where

$$a_{21} = \frac{\gamma - 1}{2}\left(\frac{q_3}{q_1}\right)^2 - \frac{3 - \gamma}{2}\left(\frac{q_2}{q_1}\right)^2$$

$$a_{41} = (\gamma - 1)\left[\left(\frac{q_2}{q_1}\right)^3 + \left(\frac{q_3}{q_1}\right)^2\left(\frac{q_2}{q_1}\right)\right] - \gamma\left(\frac{q_4}{q_1}\right)\left(\frac{q_2}{q_1}\right)$$

$$a_{42} = \gamma\left(\frac{q_4}{q_1}\right) - \frac{\gamma - 1}{2}\left[3\left(\frac{q_2}{q_1}\right)^2 + \left(\frac{q_3}{q_1}\right)^2\right]$$

$$a_{43} = -(\gamma - 1)\left(\frac{q_2}{q_1}\right)\left(\frac{q_3}{q_1}\right), \qquad (2.14)$$

and in terms of the primitive variables as

$$A = \begin{bmatrix} 0 & 1 & 0 & 0 \\ a_{21} & (3-\gamma)u & (1-\gamma)v & (\gamma-1) \\ -uv & v & u & 0 \\ a_{41} & a_{42} & a_{43} & \gamma u \end{bmatrix} , \qquad (2.15)$$

where

$$a_{21} = \frac{\gamma-1}{2}v^2 - \frac{3-\gamma}{2}u^2$$

$$a_{41} = (\gamma-1)u(u^2+v^2) - \gamma\frac{ue}{\rho}$$

$$a_{42} = \gamma\frac{e}{\rho} - \frac{\gamma-1}{2}(3u^2+v^2)$$

$$a_{43} = (1-\gamma)uv . \qquad (2.16)$$

Derivation of the two forms of $B = \partial F/\partial Q$ is similar. The eigenvalues of the flux Jacobian matrices are purely real. This is the defining feature of *hyperbolic* systems of PDE's, which are further discussed in Section 2.5. The homogeneous property of the Euler equations is discussed in Appendix C.

The Navier–Stokes equations include both convective and diffusive fluxes. This motivates the choice of our two scalar model equations associated with the physics of convection and diffusion. Furthermore, aspects of convective phenomena associated with coupled systems of equations such as the Euler equations are important in developing numerical methods and boundary conditions. Thus we also study linear hyperbolic systems of PDE's.

2.3 The Linear Convection Equation

2.3.1 Differential Form

The simplest linear model for convection and wave propagation is the linear convection equation given by the following PDE:

$$\frac{\partial u}{\partial t} + a\frac{\partial u}{\partial x} = 0 . \qquad (2.17)$$

Here $u(x,t)$ is a scalar quantity propagating with speed a, a real constant which may be positive or negative. The manner in which the boundary conditions are specified separates the following two phenomena for which this equation is a model:

(1) In one type, the scalar quantity u is given on one boundary, corresponding to a wave entering the domain through this "inflow" boundary.
No boundary condition is specified at the opposite side, the "outflow"
boundary. This is consistent in terms of the well-posedness of a first-
order PDE. Hence the wave leaves the domain through the outflow
boundary without distortion or reflection. This type of phenomenon
is referred to, simply, as the convection problem. It represents most of
the "usual" situations encountered in convecting systems. Note that
the left-hand boundary is the inflow boundary when a is positive, while
the right-hand boundary is the inflow boundary when a is negative.

(2) In the other type, the flow being simulated is periodic. At any given
time, what enters on one side of the domain must be the same as that
which is leaving on the other. This is referred to as the *biconvection*
problem. It is the simplest to study and serves to illustrate many of the
basic properties of numerical methods applied to problems involving
convection, without special consideration of boundaries. Hence, we pay
a great deal of attention to it in the initial chapters.

Now let us consider a situation in which the initial condition is given by
$u(x,0) = u_0(x)$, and the domain is infinite. It is easy to show by substitution
that the exact solution to the linear convection equation is then

$$u(x,t) = u_0(x - at) . \qquad (2.18)$$

The initial waveform propagates unaltered with speed $|a|$ to the right if a
is positive, and to the left if a is negative. With periodic boundary conditions, the waveform travels through one boundary and reappears at the other
boundary, eventually returning to its initial position. In this case, the process
continues forever without any change in the shape of the solution. Preserv-
ing the shape of the initial condition $u_0(x)$ can be a difficult challenge for a
numerical method.

2.3.2 Solution in Wave Space

We now examine the biconvection problem in more detail. Let the domain
be given by $0 \le x \le 2\pi$. We restrict our attention to initial conditions in the
form

$$u(x,0) = f(0)e^{i\kappa x} , \qquad (2.19)$$

where $f(0)$ is a complex constant, and κ is the *wavenumber*. In order to satisfy
the periodic boundary conditions, κ must be an integer. It is a measure of the
number of wavelengths within the domain. With such an initial condition,
the solution can be written as

$$u(x,t) = f(t)e^{i\kappa x} , \qquad (2.20)$$

where the time dependence is contained in the complex function $f(t)$. Substituting this solution into the linear convection equation, (2.17), we find that $f(t)$ satisfies the following ordinary differential equation (ODE)

$$\frac{\mathrm{d}f}{\mathrm{d}t} = -\mathrm{i}a\kappa f \;, \tag{2.21}$$

which has the solution

$$f(t) = f(0)\mathrm{e}^{-\mathrm{i}a\kappa t} \;. \tag{2.22}$$

Substituting $f(t)$ into (2.20) gives the following solution

$$u(x,t) = f(0)\mathrm{e}^{\mathrm{i}\kappa(x-at)} = f(0)\mathrm{e}^{\mathrm{i}(\kappa x - \omega t)} \;, \tag{2.23}$$

where the frequency, ω, the wavenumber, κ, and the phase speed, a, are related by

$$\omega = \kappa a \;. \tag{2.24}$$

The relation between the frequency and the wavenumber is known as the dispersion relation. The linear relation given by (2.24) is characteristic of wave propagation in a nondispersive medium. This means that the phase speed is the same for all wavenumbers. As we shall see later, most numerical methods introduce some dispersion; that is, in a simulation, waves with different wavenumbers travel at different speeds.

An arbitrary initial waveform can be produced by summing initial conditions of the form of (2.19). For M modes, one obtains

$$u(x,0) = \sum_{m=1}^{M} f_m(0)\mathrm{e}^{\mathrm{i}\kappa_m x} \;, \tag{2.25}$$

where the wavenumbers are often ordered such that $\kappa_1 \leq \kappa_2 \leq \cdots \leq \kappa_M$. Since the wave equation is linear, the solution is obtained by summing solutions of the form of (2.23), giving

$$u(x,t) = \sum_{m=1}^{M} f_m(0)\mathrm{e}^{\mathrm{i}\kappa_m(x-at)} \;. \tag{2.26}$$

Dispersion and dissipation resulting from a numerical approximation will cause the shape of the solution to change from that of the original waveform.

2.4 The Diffusion Equation

2.4.1 Differential Form

Diffusive fluxes are associated with molecular motion in a continuum fluid. A simple linear model equation for a diffusive process is

$$\frac{\partial u}{\partial t} = \nu \frac{\partial^2 u}{\partial x^2} , \qquad (2.27)$$

where ν is a positive real constant. For example, with u representing the temperature, this parabolic PDE governs the diffusion of heat in one dimension. Boundary conditions can be periodic, Dirichlet (specified u), Neumann (specified $\partial u/\partial x$), or mixed Dirichlet/Neumann.

In contrast to the linear convection equation, the diffusion equation has a nontrivial *steady-state* solution, which is one that satisfies the governing PDE with the partial derivative in time equal to zero. In the case of (2.27), the steady-state solution must satisfy

$$\frac{\partial^2 u}{\partial x^2} = 0 . \qquad (2.28)$$

Therefore, u must vary linearly with x at steady state such that the boundary conditions are satisfied. Other steady-state solutions are obtained if a source term $g(x)$ is added to (2.27), as follows:

$$\frac{\partial u}{\partial t} = \nu \left[\frac{\partial^2 u}{\partial x^2} - g(x) \right] , \qquad (2.29)$$

giving a steady-state solution which satisfies

$$\frac{\partial^2 u}{\partial x^2} - g(x) = 0 . \qquad (2.30)$$

In two dimensions, the diffusion equation becomes

$$\frac{\partial u}{\partial t} = \nu \left[\frac{\partial^2 u}{\partial x^2} + \frac{\partial^2 u}{\partial y^2} - g(x,y) \right] , \qquad (2.31)$$

where $g(x,y)$ is again a source term. The corresponding steady equation is

$$\frac{\partial^2 u}{\partial x^2} + \frac{\partial^2 u}{\partial y^2} - g(x,y) = 0 . \qquad (2.32)$$

While (2.31) is parabolic, (2.32) is elliptic. The latter is known as the Poisson equation for nonzero g, and as Laplace's equation for zero g.

2.4.2 Solution in Wave Space

We now consider a series solution to (2.27) with Dirichlet boundary conditions. Let the domain be given by $0 \leq x \leq \pi$ with boundary conditions $u(0) = u_a$, $u(\pi) = u_b$. It is clear that the steady-state solution is given by a linear function which satisfies the boundary conditions, i.e. $h(x) = u_a + (u_b - u_a)x/\pi$. Let the initial condition be

$$u(x,0) = \sum_{m=1}^{M} f_m(0) \sin \kappa_m x + h(x) , \tag{2.33}$$

where κ must be an integer in order to satisfy the boundary conditions. A solution of the form

$$u(x,t) = \sum_{m=1}^{M} f_m(t) \sin \kappa_m x + h(x) \tag{2.34}$$

satisfies the initial and boundary conditions. Substituting this form into (2.27) gives the following ODE for f_m:

$$\frac{\mathrm{d}f_m}{\mathrm{d}t} = -\kappa_m^2 \nu f_m \tag{2.35}$$

and we find

$$f_m(t) = f_m(0)e^{-\kappa_m^2 \nu t} . \tag{2.36}$$

Substituting $f_m(t)$ into (2.34), we obtain

$$u(x,t) = \sum_{m=1}^{M} f_m(0)e^{-\kappa_m^2 \nu t} \sin \kappa_m x + h(x) . \tag{2.37}$$

The steady-state solution ($t \to \infty$) is simply $h(x)$. Equation (2.37) shows that high wavenumber components (large κ_m) of the solution decay more rapidly than low wavenumber components, consistent with the physics of diffusion.

2.5 Linear Hyperbolic Systems

The Euler equations, (2.8), form a hyperbolic system of partial differential equations. Other systems of equations governing convection and wave-propagation phenomena, such as the Maxwell equations describing the propagation of electromagnetic waves, are also of hyperbolic type. Many aspects of numerical methods for such systems can be understood by studying a one-dimensional constant-coefficient linear system of the form

$$\frac{\partial u}{\partial t} + A\frac{\partial u}{\partial x} = 0 , \tag{2.38}$$

where $u = u(x,t)$ is a vector of length m and A is a real $m \times m$ matrix. For conservation laws, this equation can also be written in the form

$$\frac{\partial u}{\partial t} + \frac{\partial f}{\partial x} = 0 , \tag{2.39}$$

where f is the flux vector and $A = \frac{\partial f}{\partial u}$ is the flux Jacobian matrix. The entries in the flux Jacobian are

$$a_{ij} = \frac{\partial f_i}{\partial u_j} . \tag{2.40}$$

The flux Jacobian for the Euler equations is derived in Section 2.2. For a constant-coefficient linear system $f = Au$.

Such a system is hyperbolic if A is diagonalizable with real eigenvalues.[1] Thus

$$\Lambda = X^{-1}AX , \tag{2.41}$$

where Λ is a diagonal matrix containing the eigenvalues of A, and X is the matrix of right eigenvectors. Premultiplying (2.38) by X^{-1}, postmultiplying A by the product XX^{-1}, and noting that X and X^{-1} are constants, we obtain

$$\frac{\partial X^{-1}u}{\partial t} + \frac{\partial \overbrace{X^{-1}AX}^{\Lambda} X^{-1}u}{\partial x} = 0 . \tag{2.42}$$

With $w = X^{-1}u$, this can be rewritten as

$$\frac{\partial w}{\partial t} + \Lambda \frac{\partial w}{\partial x} = 0 . \tag{2.43}$$

When written in this manner, the equations have been decoupled into m scalar equations of the form

$$\frac{\partial w_i}{\partial t} + \lambda_i \frac{\partial w_i}{\partial x} = 0 . \tag{2.44}$$

The elements of w are known as *characteristic variables*. Each characteristic variable satisfies the linear convection equation with the speed given by the corresponding eigenvalue of A.

Based on the above, we see that a hyperbolic system in the form of (2.38) has a solution given by the superposition of waves which can travel in either the positive or negative directions and at varying speeds. While the scalar linear convection equation is clearly an excellent model equation for hyperbolic systems, we must ensure that our numerical methods are appropriate for wave speeds of arbitrary sign and possibly widely varying magnitudes.

The one-dimensional Euler equations can also be diagonalized, leading to three equations in the form of the linear convection equation, although they remain nonlinear, of course. The eigenvalues of the flux Jacobian matrix, or wave speeds, are $u, u + c$, and $u - c$, where u is the local fluid velocity, and $c = \sqrt{\gamma p / \rho}$ is the local speed of sound. The speed u is associated with convection of the fluid, while $u + c$ and $u - c$ are associated with sound waves. Therefore, in a supersonic flow, where $|u| > c$, all of the wave speeds have the same sign. In a subsonic flow, where $|u| < c$, wave speeds of both positive

---

[1] See Appendix A for a brief review of some basic relations and definitions from linear algebra.

and negative sign are present, corresponding to the fact that sound waves can travel upstream in a subsonic flow.

The signs of the eigenvalues of the matrix A are also important in determining suitable boundary conditions. The characteristic variables each satisfy the linear convection equation with the wave speed given by the corresponding eigenvalue. Therefore, the boundary conditions can be specified accordingly. That is, characteristic variables associated with positive eigenvalues can be specified at the left boundary, which corresponds to inflow for these variables. Characteristic variables associated with negative eigenvalues can be specified at the right boundary, which is the inflow boundary for these variables. While other boundary condition treatments are possible, they must be consistent with this approach.

Exercises

2.1 Show that the 1D Euler equations can be written in terms of the *primitive* variables $R = [\rho, u, p]^{\mathrm{T}}$ as follows:

$$\frac{\partial R}{\partial t} + M \frac{\partial R}{\partial x} = 0 \,,$$

where

$$M = \begin{bmatrix} u & \rho & 0 \\ 0 & u & \rho^{-1} \\ 0 & \gamma p & u \end{bmatrix} \,.$$

Assume an ideal gas, $p = (\gamma - 1)(e - \rho u^2/2)$.

2.2 Find the eigenvalues and eigenvectors of the matrix M derived in Exercise 2.1.

2.3 Derive the flux Jacobian matrix $A = \partial E / \partial Q$ for the 1D Euler equations resulting from the conservative variable formulation (2.5). Find its eigenvalues and compare with those obtained in Exercise 2.2.

2.4 Show that the two matrices M and A derived in Exercises 2.1 and 2.3, respectively, are related by a similarity transform. (Hint: make use of the matrix $S = \partial Q / \partial R$.)

2.5 Write the 2D diffusion equation, (2.31), in the form of (2.2).

2.6 Given the initial condition $u(x, 0) = \sin x$ defined on $0 \le x \le 2\pi$, write it in the form of (2.25), that is, find the necessary values of $f_m(0)$. (Hint: use $M = 2$ with $\kappa_1 = 1$ and $\kappa_2 = -1$.) Next consider the same initial condition defined only at $x = 2\pi j/4$, $j = 0, 1, 2, 3$. Find the values of $f_m(0)$ required to reproduce the initial condition at these discrete points using $M = 4$ with $\kappa_m = m - 1$.

2.7 Plot the first three basis functions used in constructing the exact solution to the diffusion equation in Section 2.4.2. Next consider a solution with boundary conditions $u_a = u_b = 0$, and initial conditions from (2.33) with $f_m(0) = 1$ for $1 \le m \le 3$, $f_m(0) = 0$ for $m > 3$. Plot the initial condition on the domain $0 \le x \le \pi$. Plot the solution at $t = 1$ with $\nu = 1$.

2.8 Write the classical wave equation $\partial^2 u/\partial t^2 = c^2 \partial^2 u/\partial x^2$ as a first-order system, i.e. in the form

$$\frac{\partial U}{\partial t} + A \frac{\partial U}{\partial x} = 0 \,,$$

where $U = [\partial u/\partial x, \partial u/\partial t]^{\mathrm{T}}$. Find the eigenvalues and eigenvectors of A.

2.9 The Cauchy–Riemann equations are formed from the coupling of the steady compressible continuity (conservation of mass) equation

$$\frac{\partial \rho u}{\partial x} + \frac{\partial \rho v}{\partial y} = 0$$

and the vorticity definition

$$\omega = -\frac{\partial v}{\partial x} + \frac{\partial u}{\partial y} = 0 \,,$$

where $\omega = 0$ for irrotational flow. For isentropic and homenthalpic flow, the system is closed by the relation

$$\rho = \left[1 - \frac{\gamma - 1}{2} \left(u^2 + v^2 - 1 \right) \right]^{\frac{1}{\gamma - 1}} \,.$$

Note that the variables have been non-dimensionalized. Combining the two PDE's, we have

$$\frac{\partial f(q)}{\partial x} + \frac{\partial g(q)}{\partial y} = 0 \,,$$

where

$$q = \begin{pmatrix} u \\ v \end{pmatrix}, \quad f = \begin{pmatrix} -\rho u \\ v \end{pmatrix}, \quad g = \begin{pmatrix} -\rho v \\ -u \end{pmatrix} \,.$$

One approach to solving these equations is to add a time-dependent term and find the steady solution of the following equation:

$$\frac{\partial q}{\partial t} + \frac{\partial f}{\partial x} + \frac{\partial g}{\partial y} = 0 \,.$$

(a) Find the flux Jacobians of f and g with respect to q.
(b) Determine the eigenvalues of the flux Jacobians.
(c) Determine the conditions (in terms of ρ and u) under which the system is hyperbolic, i.e., has real eigenvalues.
(d) Are the above fluxes homogeneous? (see Appendix C)

3. Finite-Difference Approximations

In common with the equations governing unsteady fluid flow, our model equations contain partial derivatives with respect to both space and time. One can approximate these simultaneously and then solve the resulting difference equations. Alternatively, one can approximate the spatial derivatives first, thereby producing a system of ordinary differential equations. The time derivatives are approximated next, leading to a *time-marching method* which produces a set of difference equations. This is the approach emphasized here. In this chapter, the concept of finite-difference approximations to partial derivatives is presented. While these can be applied either to spatial derivatives or time derivatives, our emphasis in this chapter is on spatial derivatives; time derivatives are treated in Chapter 6. Strategies for applying these finite-difference approximations will be discussed in Chapter 4.

All of the material below is presented in a Cartesian system. We emphasize the fact that quite general classes of meshes expressed in general curvilinear coordinates in *physical space* can be transformed to a uniform Cartesian mesh with equispaced intervals in a so-called *computational space*, as shown in Figure 3.1. The *computational space* is uniform; all the geometric variation is absorbed into variable coefficients of the transformed equations. For this reason, in much of the following accuracy analysis, we use an equispaced Cartesian system without being unduly restrictive or losing practical application.

3.1 Meshes and Finite-Difference Notation

The simplest mesh involving both time and space is shown in Figure 3.2. Inspection of this figure permits us to define the terms and notation needed to describe finite-difference approximations. In general, the dependent variables, u, for example, are functions of the independent variables t, and x, y, z. For the first several chapters we consider primarily the 1D case $u = u(x, t)$. When only one variable is denoted, dependence on the other is assumed. The mesh index for x is *always* j, and that for t is *always* n. Then on an equispaced grid

$$x = x_j = j\Delta x \qquad (3.1)$$

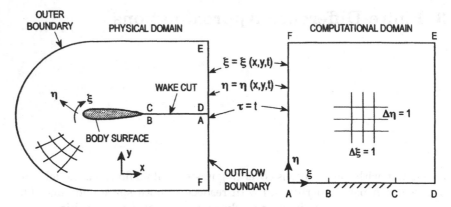

Fig. 3.1. Physical and computational spaces

$$t = t_n = n\Delta t = nh , \tag{3.2}$$

where Δx is the spacing in x and Δt the spacing in t, as shown in Figure 3.2. Note that $h = \Delta t$ throughout. Later k and l are used for y and z in a similar way. When n, j, k, l are used for other purposes (which is sometimes necessary), local context should make the meaning obvious.

The convention for subscript and superscript indexing is as follows:

$$u(t + kh) = u([n + k]h) \quad = u_{n+k}$$

$$u(x + m\Delta x) = u([j + m]\Delta x) = u_{j+m} \tag{3.3}$$

$$u(x + m\Delta x, t + kh) = u_{j+m}^{(n+k)} .$$

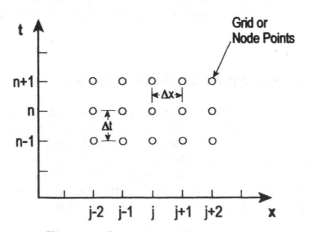

Fig. 3.2. Space–time grid arrangement

Notice that when used alone, both the time and space indices appear as a subscript, but when used together, time is always a superscript and is usually enclosed with parentheses to distinguish it from an exponent.

Derivatives are expressed according to the usual conventions. Thus for partial derivatives in space or time we use interchangeably

$$\partial_x u = \frac{\partial u}{\partial x}, \quad \partial_t u = \frac{\partial u}{\partial t}, \quad \partial_{xx} u = \frac{\partial^2 u}{\partial x^2}, \quad \text{etc.} \tag{3.4}$$

For the ordinary time derivative in the study of ODE's we use

$$u' = \frac{du}{dt} . \tag{3.5}$$

In this text, subscripts on dependent variables are never used to express derivatives. Thus u_x will *not* be used to represent the first derivative of u with respect to x.

The notation for difference approximations follows the same philosophy, but (with one exception) it is not unique. By this we mean that the symbol δ is used to represent a difference approximation to a derivative such that, for example,

$$\delta_x \approx \partial_x , \qquad \delta_{xx} \approx \partial_{xx} , \tag{3.6}$$

but the precise nature (and order) of the approximation is not carried in the symbol δ. Other ways are used to determine its precise meaning. The one exception is the symbol Δ, which is defined such that

$$\Delta t_n = t_{n+1} - t_n, \quad \Delta x_j = x_{j+1} - x_j, \quad \Delta u_n = u_{n+1} - u_n, \quad \text{etc.} \tag{3.7}$$

When there is no subscript on Δt or Δx, the spacing is uniform.

3.2 Space Derivative Approximations

A difference approximation can be generated or evaluated by means of a simple Taylor series expansion. For example, consider $u(x,t)$ with t fixed. Then, following the notation convention given in (3.1)–(3.3), $x = j\Delta x$ and $u(x + k\Delta x) = u(j\Delta x + k\Delta x) = u_{j+k}$. Expanding the latter term about x gives[1]

$$u_{j+k} = u_j + (k\Delta x)\left(\frac{\partial u}{\partial x}\right)_j + \frac{1}{2}(k\Delta x)^2\left(\frac{\partial^2 u}{\partial x^2}\right)_j + \cdots$$
$$+ \frac{1}{n!}(k\Delta x)^n\left(\frac{\partial^n u}{\partial x^n}\right)_j + \cdots . \tag{3.8}$$

---

[1] We assume that $u(x,t)$ is continuously differentiable.

Local difference approximations to a given partial derivative can be formed from linear combinations of u_j and u_{j+k} for $k = \pm 1, \pm 2, \ldots$.

For example, consider the Taylor series expansion for u_{j+1}:

$$u_{j+1} = u_j + (\Delta x)\left(\frac{\partial u}{\partial x}\right)_j + \frac{1}{2}(\Delta x)^2\left(\frac{\partial^2 u}{\partial x^2}\right)_j + \cdots$$

$$+ \frac{1}{n!}(\Delta x)^n\left(\frac{\partial^n u}{\partial x^n}\right)_j + \cdots . \tag{3.9}$$

Now subtract u_j and divide by Δx to obtain

$$\frac{u_{j+1} - u_j}{\Delta x} = \left(\frac{\partial u}{\partial x}\right)_j + \frac{1}{2}(\Delta x)\left(\frac{\partial^2 u}{\partial x^2}\right)_j + \cdots . \tag{3.10}$$

Thus the expression $(u_{j+1} - u_j)/\Delta x$ is a reasonable approximation for $\left(\frac{\partial u}{\partial x}\right)_j$ as long as Δx is small relative to some pertinent length scale. Next consider the space difference approximation $(u_{j+1} - u_{j-1})/(2\Delta x)$. Expand the terms in the numerator about j and regroup the result to form the following equation

$$\frac{u_{j+1} - u_{j-1}}{2\Delta x} - \left(\frac{\partial u}{\partial x}\right)_j = \frac{1}{6}\Delta x^2\left(\frac{\partial^3 u}{\partial x^3}\right)_j + \frac{1}{120}\Delta x^4\left(\frac{\partial^5 u}{\partial x^5}\right)_j \cdots . \tag{3.11}$$

When expressed in this manner, it is clear that the discrete terms on the left side of the equation represent a first derivative with a certain amount of error which appears on the right side of the equal sign. It is also clear that the error depends on the grid spacing to a certain order. The error term containing the grid spacing to the lowest power gives the order of the method. From (3.10), we see that the expression $(u_{j+1} - u_j)/\Delta x$ is a first-order approximation to $\left(\frac{\partial u}{\partial x}\right)_j$. Similarly, (3.11) shows that $(u_{j+1} - u_{j-1})/(2\Delta x)$ is a second-order approximation to a first derivative. The latter is referred to as the three-point centered difference approximation, and one often sees the summary result presented in the form

$$\left(\frac{\partial u}{\partial x}\right)_j = \frac{u_{j+1} - u_{j-1}}{2\Delta x} + O(\Delta x^2) . \tag{3.12}$$

3.3 Finite-Difference Operators

3.3.1 Point Difference Operators

Perhaps the most common examples of finite-difference formulas are the three-point centered-difference approximations for the first and second derivatives:[2]

---

[2] We will derive the second-derivative operator shortly.

$$\left(\frac{\partial u}{\partial x}\right)_j = \frac{1}{2\Delta x}(u_{j+1} - u_{j-1}) + O(\Delta x^2) \tag{3.13}$$

$$\left(\frac{\partial^2 u}{\partial x^2}\right)_j = \frac{1}{\Delta x^2}(u_{j+1} - 2u_j + u_{j-1}) + O(\Delta x^2) . \tag{3.14}$$

These are the basis for *point difference operators* since they give an approximation to a derivative at one discrete point in a mesh in terms of surrounding points. However, neither of these expressions tells us how other points in the mesh are differenced or how boundary conditions are enforced. Such additional information requires a more sophisticated formulation.

3.3.2 Matrix Difference Operators

Consider the relation

$$(\delta_{xx}u)_j = \frac{1}{\Delta x^2}(u_{j+1} - 2u_j + u_{j-1}) , \tag{3.15}$$

which is a point difference approximation to a second derivative. Now let us derive a *matrix* operator representation for the same approximation. Consider the four-point mesh with boundary points at a and b shown below. Notice that when we speak of "the number of points in a mesh", we mean the number of *interior* points excluding the boundaries.

	a	1	2	3	4	b
$x =$	0	–	–	–	–	π
$j =$		1	·	·	M	

Four point mesh. $\Delta x = \pi/(M+1)$

Now impose Dirichlet boundary conditions, $u(0) = u_a$, $u(\pi) = u_b$ and use the centered difference approximation given by (3.15) at every point in the mesh. We arrive at the four equations:

$$(\delta_{xx}u)_1 = \frac{1}{\Delta x^2}(u_a - 2u_1 + u_2)$$

$$(\delta_{xx}u)_2 = \frac{1}{\Delta x^2}(u_1 - 2u_2 + u_3) \tag{3.16}$$

$$(\delta_{xx}u)_3 = \frac{1}{\Delta x^2}(u_2 - 2u_3 + u_4)$$

$$(\delta_{xx}u)_4 = \frac{1}{\Delta x^2}(u_3 - 2u_4 + u_b) .$$

Writing these equations in the more suggestive form

$$\begin{aligned}
(\delta_{xx}u)_1 &= (\ u_a\ -2u_1\ +u_2 &)/\Delta x^2 \\
(\delta_{xx}u)_2 &= (\ \ \ \ \ \ u_1\ -2u_2\ +u_3 &)/\Delta x^2 \\
(\delta_{xx}u)_3 &= (\ \ \ \ \ \ \ \ \ \ \ \ u_2\ -2u_3\ +u_4 &)/\Delta x^2 \\
(\delta_{xx}u)_4 &= (\ \ \ \ \ \ \ \ \ \ \ \ \ \ \ \ \ \ u_3\ -2u_4\ +u_b &)/\Delta x^2\ ,
\end{aligned} \tag{3.17}$$

it is clear that we can express them in a vector-matrix form, and further, that the resulting matrix has a very special form. Introducing

$$\vec{u} = \begin{bmatrix} u_1 \\ u_2 \\ u_3 \\ u_4 \end{bmatrix}, \quad \left(\vec{bc}\right) = \frac{1}{\Delta x^2} \begin{bmatrix} u_a \\ 0 \\ 0 \\ u_b \end{bmatrix} \tag{3.18}$$

and

$$A = \frac{1}{\Delta x^2} \begin{bmatrix} -2 & 1 & & \\ 1 & -2 & 1 & \\ & 1 & -2 & 1 \\ & & 1 & -2 \end{bmatrix}, \tag{3.19}$$

we can rewrite (3.17) as

$$\delta_{xx}\vec{u} = A\vec{u} + \left(\vec{bc}\right). \tag{3.20}$$

This example illustrates a matrix difference operator. Each line of a matrix difference operator is based on a point difference operator, but the point operators used from line to line are not necessarily the same. For example, boundary conditions may dictate that the lines at or near the bottom or top of the matrix be modified. In the extreme case of the matrix difference operator representing a spectral method, none of the lines is the same. The matrix operators representing the three-point central-difference approximations for a first and second derivative with Dirichlet boundary conditions on a four-point mesh are

$$\delta_x = \frac{1}{2\Delta x} \begin{bmatrix} 0 & 1 & & \\ -1 & 0 & 1 & \\ & -1 & 0 & 1 \\ & & -1 & 0 \end{bmatrix}, \quad \delta_{xx} = \frac{1}{\Delta x^2} \begin{bmatrix} -2 & 1 & & \\ 1 & -2 & 1 & \\ & 1 & -2 & 1 \\ & & 1 & -2 \end{bmatrix}. \tag{3.21}$$

As a further example, replace the fourth line in (3.16) by the following point operator for a Neumann boundary condition (see Section 3.6):

$$(\delta_{xx}u)_4 = \frac{2}{3}\frac{1}{\Delta x}\left(\frac{\partial u}{\partial x}\right)_b - \frac{2}{3}\frac{1}{\Delta x^2}(u_4 - u_3), \tag{3.22}$$

where the boundary condition is

$$\left(\frac{\partial u}{\partial x}\right)_{x=\pi} = \left(\frac{\partial u}{\partial x}\right)_b. \tag{3.23}$$

Then the matrix operator for a three-point central-differencing scheme at interior points and *a second-order approximation* for a Neumann condition on the right is given by

$$\delta_{xx} = \frac{1}{\Delta x^2} \begin{bmatrix} -2 & 1 & & \\ 1 & -2 & 1 & \\ & 1 & -2 & 1 \\ & & 2/3 & -2/3 \end{bmatrix}. \tag{3.24}$$

Each of these matrix difference operators is a square matrix with elements that are all zeros except for those along bands which are clustered around the central diagonal. We call such a matrix a *banded matrix* and introduce the notation

$$B(M : a, b, c) = \begin{bmatrix} b & c & & & \\ a & b & c & & \\ & & \ddots & & \\ & & a & b & c \\ & & & a & b \end{bmatrix} \begin{matrix} 1 \\ \\ \vdots \\ \\ M \end{matrix}, \tag{3.25}$$

where the matrix dimensions are $M \times M$. Use of M in the argument is optional, and the illustration is given for a simple *tridiagonal* matrix although any number of bands is a possibility. A tridiagonal matrix without constants along the bands can be expressed as $B(\vec{a}, \vec{b}, \vec{c})$. The arguments for a banded matrix are always odd in number, and the central one *always* refers to the central diagonal.

We can now generalize our previous examples. Defining $\vec{u}$ as[3]

$$\vec{u} = \begin{bmatrix} u_1 \\ u_2 \\ u_3 \\ \vdots \\ u_M \end{bmatrix}, \tag{3.26}$$

we can approximate the second derivative of $\vec{u}$ by

$$\delta_{xx}\vec{u} = \frac{1}{\Delta x^2} B(1, -2, 1)\vec{u} + \left(\vec{bc}\right), \tag{3.27}$$

where $\left(\vec{bc}\right)$ stands for the vector holding the Dirichlet boundary conditions on the left and right sides:

$$\left(\vec{bc}\right) = \frac{1}{\Delta x^2}[u_a, 0, \cdots, 0, u_b]^{\mathrm{T}}. \tag{3.28}$$

---

[3] Note that $\vec{u}$ is a function of time only, since each element corresponds to one specific spatial location.

If we prescribe Neumann boundary conditions on the right side, as in (3.24), we find

$$\delta_{xx}\vec{u} = \frac{1}{\Delta x^2}B(\vec{a},\vec{b},1)\vec{u} + \left(\vec{bc}\right) , \qquad (3.29)$$

where

$$\vec{a} = [1,1,\cdots,2/3]^T$$

$$\vec{b} = [-2,-2,-2,\cdots,-2/3]^T$$

$$\left(\vec{bc}\right) = \frac{1}{\Delta x^2}\left[u_a,0,0,\cdots,\frac{2\Delta x}{3}\left(\frac{\partial u}{\partial x}\right)_b\right]^T .$$

Notice that the matrix operators given by (3.27) and (3.29) carry more information than the point operator given by (3.15). In (3.27) and (3.29), the boundary conditions have been uniquely specified and it is clear that the same point operator has been applied at every point in the field except at the boundaries. The ability to specify in the matrix derivative operator the exact nature of the approximation at the various points in the field including the boundaries permits the use of quite general constructions which will be useful later in considerations of stability.

Since we make considerable use of both matrix and point operators, it is important to establish a relation between them. A point operator is generally written for some derivative at the reference point j in terms of neighboring values of the function. For example

$$(\delta_x u)_j = a_2 u_{j-2} + a_1 u_{j-1} + bu_j + c_1 u_{j+1} \qquad (3.30)$$

might be the point operator for a first derivative. The corresponding matrix operator has for its arguments the coefficients giving the weights to the values of the function at the various locations. A j-shift in the point operator corresponds to a diagonal shift in the matrix operator. Thus the matrix equivalent of (3.30) is

$$\delta_x \vec{u} = B(a_2,a_1,b,c_1,0)\vec{u} . \qquad (3.31)$$

Note the addition of a zero in the fifth element which makes it clear that b is the coefficient of u_j.

3.3.3 Periodic Matrices

The above illustrated cases in which the boundary conditions are fixed. If the boundary conditions are *periodic*, the form of the matrix operator changes. Consider the eight-point periodic mesh shown below. This can either be presented on a linear mesh with repeated entries, or more suggestively on a

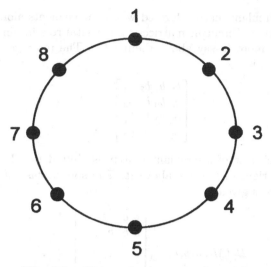

Fig. 3.3. Eight points on a circular mesh

circular mesh as in Figure 3.3. When the mesh is laid out on the perimeter of
a circle, it does not matter where the numbering starts, as long as it "ends"
at the point just preceding its starting location.

		7	8	1	2	3	4	5	6	7	8	1	2	
$\cdots$														$\cdots$
$x =$		$-$	$-$	0	$-$	$-$	$-$	$-$	$-$	$-$	$-$	2π	$-$	
$j =$				0	1	$\cdot$	$\cdot$	$\cdot$	$\cdot$	$\cdot$	$\cdot$	M		

Eight points on a linear periodic mesh. $\Delta x = 2\pi/M$

The matrix that represents differencing schemes for scalar equations on a
periodic mesh is referred to as a *periodic* matrix. A typical periodic tridiagonal
matrix operator with nonuniform entries is given for a 6-point mesh by

$$B_{\mathrm{p}}(6:\vec{a},\vec{b},\vec{c}) = \begin{bmatrix} b_1 & c_2 & & & & a_6 \\ a_1 & b_2 & c_3 & & & \\ & a_2 & b_3 & c_4 & & \\ & & a_3 & b_4 & c_5 & \\ & & & a_4 & b_5 & c_6 \\ c_1 & & & & a_5 & b_6 \end{bmatrix} . \tag{3.32}$$

3.3.4 Circulant Matrices

In general, as shown in the example, the elements along the diagonals of
a periodic matrix are not constant. However, a special subset of a periodic

matrix is the circulant matrix, formed when the elements along the various
bands are constant. Circulant matrices play a vital role in our analysis. We
will have much more to say about them later. The most general circulant
matrix of order 4 is

$$
\begin{bmatrix}
b_0 & b_1 & b_2 & b_3 \\
b_3 & b_0 & b_1 & b_2 \\
b_2 & b_3 & b_0 & b_1 \\
b_1 & b_2 & b_3 & b_0
\end{bmatrix} .
\tag{3.33}
$$

Notice that each row of a circulant matrix is shifted (see Figure 3.3) one
element to the right of the one above it. The special case of a tridiagonal
circulant matrix is given by

$$
B_p(M : a, b, c) =
\begin{bmatrix}
b & c & & a \\
a & b & c & \\
& & \ddots & \\
& & a & b & c \\
c & & & a & b
\end{bmatrix}
\begin{matrix} 1 \\ \\ \\ \vdots \\ \\ M \end{matrix} \cdot
\tag{3.34}
$$

When the standard three-point central-differencing approximations for a
first and second derivative (see (3.21)) are used with periodic boundary conditions, they take the form

$$
(\delta_x)_p = \frac{1}{2\Delta x}
\begin{bmatrix}
0 & 1 & & -1 \\
-1 & 0 & 1 & \\
& -1 & 0 & 1 \\
1 & & -1 & 0
\end{bmatrix}
= \frac{1}{2\Delta x} B_p(-1,0,1)
$$

and

$$
(\delta_{xx})_p = \frac{1}{\Delta x^2}
\begin{bmatrix}
-2 & 1 & & 1 \\
1 & -2 & 1 & \\
& 1 & -2 & 1 \\
1 & & 1 & -2
\end{bmatrix}
= \frac{1}{\Delta x^2} B_p(1,-2,1) .
\tag{3.35}
$$

Clearly, these special cases of periodic operators are also circulant operators.
Later on we take advantage of this special property. Notice that there are
no boundary condition vectors since this information is all interior to the
matrices themselves.

3.4 Constructing Differencing Schemes of Any Order

3.4.1 Taylor Tables

The Taylor series expansion of functions about a fixed point provides a means
for constructing finite-difference point operators of any order. A simple and

straightforward way to carry this out is to construct a "Taylor table", which makes extensive use of the expansion given by (3.8). As an example, consider Table 3.1, which represents a Taylor table for an approximation of a second derivative using three values of the function centered about the point at which the derivative is to be evaluated.

The table is constructed so that some of the algebra is simplified. At the top of the table we see an expression with a question mark. This represents one of the questions that a study of this table can answer; namely, what is the local error caused by the use of this approximation? Notice that all of the terms in the equation appear in a column at the left of the table (although, in this case, Δx^2 has been multiplied into each term in order to simplify the terms to be put into the table). Then notice that at the head of each column there appears the common factor that occurs in the expansion of each term about the point j, that is,

$$\Delta x^k \left(\frac{\partial^k u}{\partial x^k} \right)_j, \quad k = 0, 1, 2, \dots \ .$$

The columns to the right of the leftmost one, under the headings, make up the Taylor table. Each entry is the coefficient of the term at the top of the corresponding column in the Taylor series expansion of the term to the left of the corresponding row. For example, the last row in the table corresponds to the Taylor series expansion of $-c\,u_{j+1}$:

$$-c\,u_{j+1} = -c\,u_j - c \cdot (1) \cdot \frac{1}{1}\Delta x \left(\frac{\partial u}{\partial x} \right)_j - c \cdot (1)^2 \cdot \frac{1}{2}\Delta x^2 \left(\frac{\partial^2 u}{\partial x^2} \right)_j$$

$$-c \cdot (1)^3 \cdot \frac{1}{6}\Delta x^3 \left(\frac{\partial^3 u}{\partial x^3} \right)_j - c \cdot (1)^4 \cdot \frac{1}{24}\Delta x^4 \left(\frac{\partial^4 u}{\partial x^4} \right)_j - \cdots \ .$$

$$(3.36)$$

A Taylor table is simply a convenient way of forming linear combinations of Taylor series on a term by term basis.

Consider the sum of each of these columns. To maximize the order of accuracy of the method, we proceed from left to right and force, by the proper choice of a, b, and c, these sums to be zero. One can easily show that the sums of the first three columns are zero if we satisfy the equation

$$\begin{bmatrix} -1 & -1 & -1 \\ 1 & 0 & -1 \\ -1 & 0 & -1 \end{bmatrix} \begin{bmatrix} a \\ b \\ c \end{bmatrix} = \begin{bmatrix} 0 \\ 0 \\ -2 \end{bmatrix} \ .$$

The solution is given by $[a, b, c] = [1, -2, 1]$.

The columns that do not sum to zero constitute the error.

We designate the first non-vanishing sum to be er_t, and refer to it as the Taylor series error.

Table 3.1. Taylor table for centered 3-point Lagrangian approximation to a second derivative

$$\left(\frac{\partial^2 u}{\partial x^2}\right)_j - \frac{1}{\Delta x^2}\left(a\,u_{j-1} + b\,u_j + c\,u_{j+1}\right) = ?$$

	u_j	$\Delta x \left(\frac{\partial u}{\partial x}\right)_j$	$\Delta x^2 \left(\frac{\partial^2 u}{\partial x^2}\right)_j$	$\Delta x^3 \left(\frac{\partial^3 u}{\partial x^3}\right)_j$	$\Delta x^4 \left(\frac{\partial^4 u}{\partial x^4}\right)_j$
$\Delta x^2 \left(\frac{\partial^2 u}{\partial x^2}\right)_j$			1		
$-a \cdot u_{j-1}$	$-a$	$-a\cdot(-1)\cdot\frac{1}{1}$	$-a\cdot(-1)^2\cdot\frac{1}{2}$	$-a\cdot(-1)^3\cdot\frac{1}{6}$	$-a\cdot(-1)^4\cdot\frac{1}{24}$
$-b \cdot u_j$	$-b$				
$-c \cdot u_{j+1}$	$-c$	$-c\cdot(1)\cdot\frac{1}{1}$	$-c\cdot(1)^2\cdot\frac{1}{2}$	$-c\cdot(1)^3\cdot\frac{1}{6}$	$-c\cdot(1)^4\cdot\frac{1}{24}$

In this case er_t occurs at the fifth column in the table (for this example all even columns will vanish by symmetry) and one finds

$$er_t = \frac{1}{\Delta x^2}\left[\frac{-a}{24} + \frac{-c}{24}\right]\Delta x^4\left(\frac{\partial^4 u}{\partial x^4}\right)_j = \frac{-\Delta x^2}{12}\left(\frac{\partial^4 u}{\partial x^4}\right)_j . \qquad (3.37)$$

Note that Δx^2 has been divided through to make the error term consistent. We have just derived the familiar 3-point central-differencing point operator for a second derivative

$$\left(\frac{\partial^2 u}{\partial x^2}\right)_j - \frac{1}{\Delta x^2}(u_{j-1} - 2u_j + u_{j+1}) = O(\Delta x^2) . \qquad (3.38)$$

The Taylor table for a 3-point backward-differencing operator representing a first derivative is shown in Table 3.2. This time the first three columns sum to zero if

$$\begin{bmatrix} -1 & -1 & -1 \\ 2 & 1 & 0 \\ -4 & -1 & 0 \end{bmatrix}\begin{bmatrix} a_2 \\ a_1 \\ b \end{bmatrix} = \begin{bmatrix} 0 \\ -1 \\ 0 \end{bmatrix} ,$$

which gives $[a_2, a_1, b] = \frac{1}{2}[1, -4, 3]$. In this case the fourth column provides the leading truncation error term:

$$er_t = \frac{1}{\Delta x}\left[\frac{8a_2}{6} + \frac{a_1}{6}\right]\Delta x^3\left(\frac{\partial^3 u}{\partial x^3}\right)_j = \frac{\Delta x^2}{3}\left(\frac{\partial^3 u}{\partial x^3}\right)_j . \qquad (3.39)$$

Thus we have derived a second-order backward-difference approximation of a first derivative:

$$\left(\frac{\partial u}{\partial x}\right)_j - \frac{1}{2\Delta x}(u_{j-2} - 4u_{j-1} + 3u_j) = O(\Delta x^2) . \qquad (3.40)$$

3.4.2 Generalization of Difference Formulas

In general, a difference approximation to the mth derivative at grid point j can be cast in terms of $q + p + 1$ neighboring points as

$$\left(\frac{\partial^m u}{\partial x^m}\right)_j - \sum_{i=-p}^{q} a_i u_{j+i} = er_t , \qquad (3.41)$$

where the a_i are coefficients to be determined through the use of Taylor tables to produce approximations of a given order. Clearly this process can be used to find forward, backward, skewed, or central point operators of any order for any derivative. It could be computer automated and extended to higher dimensions. More important, however, is the fact that it can be further generalized. In order to do this, let us approach the subject in a slightly different way, that is from the point of view of interpolation formulas. These formulas are discussed in many texts on numerical analysis.

Table 3.2. Taylor table for backward 3-point Lagrangian approximation to a first derivative

$$\left(\frac{\partial u}{\partial x}\right)_j - \frac{1}{\Delta x}\left(a_2 u_{j-2} + a_1 u_{j-1} + b\, u_j\right) = ?$$

	u_j	$\Delta x \left(\frac{\partial u}{\partial x}\right)_j$	$\Delta x^2 \left(\frac{\partial^2 u}{\partial x^2}\right)_j$	$\Delta x^3 \left(\frac{\partial^3 u}{\partial x^3}\right)_j$	$\Delta x^4 \left(\frac{\partial^4 u}{\partial x^4}\right)_j$
$\Delta x \left(\frac{\partial u}{\partial x}\right)_j$		1			
$-a_2 \cdot u_{j-2}$	$-a_2$	$-a_2\cdot(-2)\cdot\frac{1}{1}$	$-a_2\cdot(-2)^2\cdot\frac{1}{2}$	$-a_2\cdot(-2)^3\cdot\frac{1}{6}$	$-a_2\cdot(-2)^4\cdot\frac{1}{24}$
$-a_1 \cdot u_{j-1}$	$-a_1$	$-a_1\cdot(-1)\cdot\frac{1}{1}$	$-a_1\cdot(-1)^2\cdot\frac{1}{2}$	$-a_1\cdot(-1)^3\cdot\frac{1}{6}$	$-a_1\cdot(-1)^4\cdot\frac{1}{24}$
$-b \cdot u_j$	$-b$				

3.4.3 Lagrange and Hermite Interpolation Polynomials

The Lagrangian interpolation polynomial is given by

$$u(x) = \sum_{k=0}^{K} a_k(x) u_k ,$$ (3.42)

where $a_k(x)$ are polynomials in x of degree K. The construction of the $a_k(x)$ can be taken from the simple Lagrangian formula for quadratic interpolation (or extrapolation) with non-equispaced points

$$u(x) = u_0 \frac{(x_1 - x)(x_2 - x)}{(x_1 - x_0)(x_2 - x_0)} + u_1 \frac{(x_0 - x)(x_2 - x)}{(x_0 - x_1)(x_2 - x_1)}$$
$$+ u_2 \frac{(x_0 - x)(x_1 - x)}{(x_0 - x_2)(x_1 - x_2)} .$$ (3.43)

Notice that the coefficient of each u_k is one when $x = x_k$, and zero when x takes any other discrete value in the set. If we take the first or second derivative of $u(x)$, impose an equispaced mesh, and evaluate these derivatives at the appropriate discrete point, we rederive the finite-difference approximations just presented. Finite-difference schemes that can be derived from (3.42) are referred to as Lagrangian approximations.

A generalization of the Lagrangian approach is brought about by using Hermitian interpolation. To construct a polynomial for $u(x)$, Hermite formulas use values of the function *and its derivative(s)* at given points in space. Our illustration is for the case in which discrete values of the function and its first derivative are used, producing the expression

$$u(x) = \sum a_k(x) u_k + \sum b_k(x) \left(\frac{\partial u}{\partial x} \right)_k .$$ (3.44)

Obviously higher-order derivatives could be included as the problems dictate. A complete discussion of these polynomials can be found in many references on numerical methods, but here we need only the concept.

The previous examples of a Taylor table constructed *explicit* point difference operators from Lagrangian interpolation formulas. Consider next the Taylor table for an *implicit* space differencing scheme for a first derivative arising from the use of an Hermite interpolation formula. A generalization of (3.41) can include derivatives at neighboring points, i.e.

$$\sum_{i=-r}^{s} b_i \left(\frac{\partial^m u}{\partial x^m} \right)_{j+i} - \sum_{i=-p}^{q} a_i u_{j+i} = er_t ,$$ (3.45)

analogous to (3.44). An example formula is illustrated at the top of Table 3.3. Here not only is the derivative at point j represented, but also included

Table 3.3. Taylor table for central 3-point Hermitian approximation to a first derivative

$$d\left(\frac{\partial u}{\partial x}\right)_{j-1} + \left(\frac{\partial u}{\partial x}\right)_{j} + e\left(\frac{\partial u}{\partial x}\right)_{j+1} - \frac{1}{\Delta x}\left(au_{j-1} + bu_j + cu_{j+1}\right) = ?$$

	u_j	$\Delta x\left(\frac{\partial u}{\partial x}\right)_j$	$\Delta x^2\left(\frac{\partial^2 u}{\partial x^2}\right)_j$	$\Delta x^3\left(\frac{\partial^3 u}{\partial x^3}\right)_j$	$\Delta x^4\left(\frac{\partial^4 u}{\partial x^4}\right)_j$	$\Delta x^5\left(\frac{\partial^5 u}{\partial x^5}\right)_j$
$\Delta x\, d\left(\frac{\partial u}{\partial x}\right)_{j-1}$		d	$d\cdot(-1)\cdot\frac{1}{1}$	$d\cdot(-1)^2\cdot\frac{1}{2}$	$d\cdot(-1)^3\cdot\frac{1}{6}$	$d\cdot(-1)^4\cdot\frac{1}{24}$
$\Delta x\left(\frac{\partial u}{\partial x}\right)_j$		1				
$\Delta x\, e\left(\frac{\partial u}{\partial x}\right)_{j+1}$		e	$e\cdot(1)\cdot\frac{1}{1}$	$e\cdot(1)^2\cdot\frac{1}{2}$	$e\cdot(1)^3\cdot\frac{1}{6}$	$e\cdot(1)^4\cdot\frac{1}{24}$
$-a\cdot u_{j-1}$	$-a$	$-a\cdot(-1)\cdot\frac{1}{1}$	$-a\cdot(-1)^2\cdot\frac{1}{2}$	$-a\cdot(-1)^3\cdot\frac{1}{6}$	$-a\cdot(-1)^4\cdot\frac{1}{24}$	$-a\cdot(-1)^5\cdot\frac{1}{120}$
$-b\cdot u_j$	$-b$					
$-c\cdot u_{j+1}$	$-c$	$-c\cdot(1)\cdot\frac{1}{1}$	$-c\cdot(1)^2\cdot\frac{1}{2}$	$-c\cdot(1)^3\cdot\frac{1}{6}$	$-c\cdot(1)^4\cdot\frac{1}{24}$	$-c\cdot(1)^5\cdot\frac{1}{120}$

are derivatives at points $j-1$ and $j+1$, which also must be expanded using Taylor series about point j. This requires the following generalization of the Taylor series expansion given in (3.8):

$$\left(\frac{\partial^m u}{\partial x^m}\right)_{j+k} = \left\{\left[\sum_{n=0}^{\infty}\frac{1}{n!}(k\Delta x)^n\frac{\partial^n}{\partial x^n}\right]\left(\frac{\partial^m u}{\partial x^m}\right)\right\}_j. \qquad (3.46)$$

The derivative terms now have coefficients (the coefficient on the j point is taken as one to simplify the algebra) which must be determined using the Taylor table approach as outlined below.

To maximize the order of accuracy, we must satisfy the relation

$$\begin{bmatrix} -1 & -1 & -1 & 0 & 0 \\ 1 & 0 & -1 & 1 & 1 \\ -1 & 0 & -1 & -2 & 2 \\ 1 & 0 & -1 & 3 & 3 \\ -1 & 0 & -1 & -4 & 4 \end{bmatrix}\begin{bmatrix} a \\ b \\ c \\ d \\ e \end{bmatrix} = \begin{bmatrix} 0 \\ -1 \\ 0 \\ 0 \\ 0 \end{bmatrix}$$

having the solution $[a, b, c, d, e] = \frac{1}{4}[-3, 0, 3, 1, 1]$. Under these conditions, the sixth column sums to

$$er_t = \frac{\Delta x^4}{120}\left(\frac{\partial^5 u}{\partial x^5}\right)_j \qquad (3.47)$$

and the method can be expressed as

$$\left(\frac{\partial u}{\partial x}\right)_{j-1} + 4\left(\frac{\partial u}{\partial x}\right)_j + \left(\frac{\partial u}{\partial x}\right)_{j+1} - \frac{3}{\Delta x}(-u_{j-1} + u_{j+1}) = O(\Delta x^4). \qquad (3.48)$$

This is also referred to as a *Padé* or *compact* formula.

3.4.4 Practical Application of Padé Formulas

It is one thing to construct methods using the Hermitian concept and quite another to implement them in a computer code. In the form of a point operator it is probably not evident at first just how (3.48) can be applied. However, the situation is quite easy to comprehend if we express the same method in the form of a matrix operator. A banded matrix notation for (3.48) is

$$\frac{1}{6}B(1,4,1)\delta_x\vec{u} = \frac{1}{2\Delta x}B(-1,0,1)\vec{u} + \left(\vec{bc}\right), \qquad (3.49)$$

in which Dirichlet boundary conditions have been imposed.[4] Mathematically this is equivalent to

---

[4] In this case the vector containing the boundary conditions would include values of both u and $\partial u/\partial x$ at both boundaries.

$$\delta_x \vec{u} = 6[B(1,4,1)]^{-1}\left[\frac{1}{2\Delta x}B(-1,0,1)\vec{u} + \left(\vec{bc}\right)\right], \qquad (3.50)$$

which can be reexpressed by the "predictor–corrector" sequence

$$\vec{\tilde{u}} \;\; = \frac{1}{2\Delta x}B(-1,0,1)\vec{u} + \left(\vec{bc}\right) \qquad (3.51)$$

$$\delta_x \vec{u} = 6\,[B(1,4,1)]^{-1}\vec{\tilde{u}}\;.$$

With respect to practical implementation, the meaning of the predictor in this sequence should be clear. It simply says: take the vector array $\vec{u}$, difference it, add the boundary conditions, and store the result in the intermediate array $\vec{\tilde{u}}$. The meaning of the second row is more subtle, since it is demanding the evaluation of an inverse operator, but it still can be given a simple interpretation. An inverse matrix operator implies the solution of a coupled set of linear equations. These operators are very common in finite difference applications. They appear in the form of banded matrices having a small bandwidth, in this case a tridiagonal. The evaluation of $[B(1,4,1)]^{-1}$ is found by means of a tridiagonal "solver", which is simple to code, efficient to run, and widely used. In general, Hermitian or Padé approximations can be practical when they can be implemented by the use of efficient banded solvers.

3.4.5 Other Higher-Order Schemes

Hermitian forms of the second derivative can also be easily derived by means of a Taylor table. For example

$$\delta_{xx}\vec{u} = 12\,[B(1,10,1)]^{-1}\left[\frac{1}{\Delta x^2}B(1,-2,1)\vec{u} + \left(\vec{bc}\right)\right] \qquad (3.52)$$

is $O(\Delta x^4)$ and makes use of only tridiagonal operations. It should be mentioned that the spline approximation is one form of a Padé matrix difference operator. It is given by

$$\delta_{xx}\vec{u} = 6\,[B(1,4,1)]^{-1}\left[\frac{1}{\Delta x^2}B(1,-2,1)\vec{u} + \left(\vec{bc}\right)\right] \qquad (3.53)$$

but its order of accuracy is only $O(\Delta x^2)$. How much this reduction in accuracy is offset by the increased global continuity built into a spline fit is not known. We note that the spline fit of a first derivative is identical to any of the expressions in (3.48)–(3.51).

A final word on Hermitian approximations. Clearly they have an advantage over 3-point Lagrangian schemes because of their increased accuracy. However, a more subtle point is that they get this increase in accuracy using information that is still local to the point where the derivatives are being evaluated. In application, this can be advantageous at boundaries and in the vicinity of steep gradients. It is obvious, of course, that five-point schemes using Lagrangian approximations can be derived that have the same order of accuracy as the methods given in (3.48) and (3.52), but they will have a wider spread of space indices. In particular, two Lagrangian schemes with the same order of accuracy are (here we ignore the problem created by the boundary conditions, although this is one of the principal issues in applying these schemes)

$$\frac{\partial \vec{u}}{\partial x} - \frac{1}{12\Delta x}B_{\mathrm{p}}(1,-8,0,8,-1)\vec{u} = O(\Delta x^4) \qquad (3.54)$$

$$\frac{\partial^2 \vec{u}}{\partial x^2} - \frac{1}{12\Delta x^2}B_{\mathrm{p}}(-1,16,-30,16,-1)\vec{u} = O(\Delta x^4) . \qquad (3.55)$$

3.5 Fourier Error Analysis

In order to select a finite-difference scheme for a given application one must be able to assess the accuracy of the candidate schemes. The accuracy of an operator is often expressed in terms of the order of the leading error term determined from a Taylor table. While this is a useful measure, it provides a fairly limited description. Further information about the error behavior of a finite-difference scheme can be obtained using Fourier error analysis.

3.5.1 Application to a Spatial Operator

An arbitrary periodic function can be decomposed into its Fourier components, which are in the form $e^{i\kappa x}$, where κ is the wavenumber. It is therefore of interest to examine how well a given finite-difference operator approximates derivatives of $e^{i\kappa x}$. We will concentrate here on first derivative approximations, although the analysis is equally applicable to higher derivatives.

The exact first derivative of $e^{i\kappa x}$ is

$$\frac{\partial e^{i\kappa x}}{\partial x} = i\kappa e^{i\kappa x} . \qquad (3.56)$$

If we apply, for example, a second-order centered difference operator to $u_j = e^{i\kappa x_j}$, where $x_j = j\Delta x$, we get

$$(\delta_x u)_j = \frac{u_{j+1} - u_{j-1}}{2\Delta x}$$

$$= \frac{e^{i\kappa \Delta x(j+1)} - e^{i\kappa \Delta x(j-1)}}{2\Delta x}$$

$$= \frac{(e^{i\kappa\Delta x} - e^{-i\kappa\Delta x})e^{i\kappa x_j}}{2\Delta x}$$

$$= \frac{1}{2\Delta x}\left[(\cos\kappa\Delta x + i\sin\kappa\Delta x) - (\cos\kappa\Delta x - i\sin\kappa\Delta x)\right]e^{i\kappa x_j}$$

$$= i\frac{\sin\kappa\Delta x}{\Delta x}e^{i\kappa x_j}$$

$$= i\kappa^* e^{i\kappa x_j} , \tag{3.57}$$

where κ^* is the modified wavenumber. The modified wavenumber is so named because it appears where the wavenumber, κ, appears in the exact expression. Thus the degree to which the modified wavenumber approximates the actual wavenumber is a measure of the accuracy of the approximation.

For the second-order centered difference operator the modified wavenumber is given by

$$\kappa^* = \frac{\sin\kappa\Delta x}{\Delta x} . \tag{3.58}$$

Note that κ^* approximates κ to second-order accuracy, as is to be expected, since

$$\frac{\sin\kappa\Delta x}{\Delta x} = \kappa - \frac{\kappa^3\Delta x^2}{6} + \cdots .$$

Equation (3.58) is plotted in Figure 3.4, along with similar relations for the standard fourth-order centered difference scheme and the fourth-order Padé scheme. The expression for the modified wavenumber provides the accuracy with which a given wavenumber component of the solution is resolved for the entire wavenumber range available in a mesh of a given size, $0 \leq \kappa\Delta x \leq \pi$.

In general, finite-difference operators can be written in the form

$$(\delta_x)_j = (\delta_x^a)_j + (\delta_x^s)_j ,$$

where $(\delta_x^a)_j$ is an antisymmetric operator and $(\delta_x^s)_j$ is a symmetric operator.[5] If we restrict our interest to schemes extending from $j-3$ to $j+3$, then

$$(\delta_x^a u)_j = \frac{1}{\Delta x}[a_1(u_{j+1} - u_{j-1}) + a_2(u_{j+2} - u_{j-2}) + a_3(u_{j+3} - u_{j-3})]$$

and

$$(\delta_x^s u)_j = \frac{1}{\Delta x}[d_0 u_j + d_1(u_{j+1} + u_{j-1}) + d_2(u_{j+2} + u_{j-2}) + d_3(u_{j+3} + u_{j-3})] .$$

The corresponding modified wavenumber is

[5] In terms of a circulant matrix operator A, the antisymmetric part is obtained from $(A - A^T)/2$ and the symmetric part from $(A + A^T)/2$.

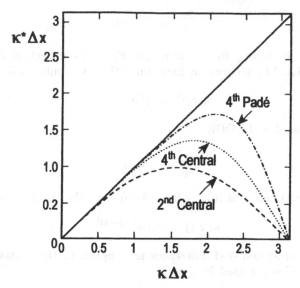

Fig. 3.4. Modified wavenumber for various schemes

$$i\kappa^* = \frac{1}{\Delta x}[d_0 + 2(d_1 \cos \kappa\Delta x + d_2 \cos 2\kappa\Delta x + d_3 \cos 3\kappa\Delta x)$$
$$+ 2i(a_1 \sin \kappa\Delta x + a_2 \sin 2\kappa\Delta x + a_3 \sin 3\kappa\Delta x)] \ . \qquad (3.59)$$

When the finite-difference operator is antisymmetric (centered), the modified wavenumber is purely real. When the operator includes a symmetric component, the modified wavenumber is complex, with the imaginary component being entirely error. The fourth-order Padé scheme is given by

$$(\delta_x u)_{j-1} + 4(\delta_x u)_j + (\delta_x u)_{j+1} = \frac{3}{\Delta x}(u_{j+1} - u_{j-1}) \ .$$

The modified wavenumber for this scheme satisfies[6]

$$i\kappa^* e^{-i\kappa\Delta x} + 4i\kappa^* + i\kappa^* e^{i\kappa\Delta x} = \frac{3}{\Delta x}(e^{i\kappa\Delta x} - e^{-i\kappa\Delta x}) \ ,$$

which gives

$$i\kappa^* = \frac{3i \sin \kappa\Delta x}{(2 + \cos \kappa\Delta x)\Delta x} \ .$$

The modified wavenumber provides a useful tool for assessing difference approximations. In the context of the linear convection equation, the errors can be given a physical interpretation. Consider once again the linear convection equation in the form

---

[6] Note that terms such as $(\delta_x u)_{j-1}$ are handled by letting $(\delta_x u)_j = i\kappa^* e^{i\kappa j \Delta x}$ and evaluating the shift in j.

$$\frac{\partial u}{\partial t} + a\frac{\partial u}{\partial x} = 0$$

on a domain extending from $-\infty$ to ∞. Recall from Section 2.3.2 that a solution initiated by a harmonic function with wavenumber κ is

$$u(x,t) = f(t)e^{i\kappa x} , \qquad (3.60)$$

where $f(t)$ satisfies the ODE

$$\frac{df}{dt} = -ia\kappa f .$$

Solving for $f(t)$ and substituting into (3.60) gives the exact solution as

$$u(x,t) = f(0)e^{i\kappa(x-at)} .$$

If second-order centered differences are applied to the spatial term, the following ODE is obtained for $f(t)$:

$$\frac{df}{dt} = -ia\left(\frac{\sin\kappa\Delta x}{\Delta x}\right)f = -ia\kappa^* f . \qquad (3.61)$$

Solving this ODE exactly (since we are considering the error from the spatial approximation only) and substituting into (3.60), we obtain

$$u_{\text{numerical}}(x,t) = f(0)e^{i\kappa(x-a^*t)} , \qquad (3.62)$$

where a^* is the numerical (or modified) phase speed, which is related to the modified wavenumber by

$$\frac{a^*}{a} = \frac{\kappa^*}{\kappa} .$$

For the above example,

$$\frac{a^*}{a} = \frac{\sin\kappa\Delta x}{\kappa\Delta x} .$$

The numerical phase speed is the speed at which a harmonic function is propagated numerically. Since $a^*/a \leq 1$ for this example, the numerical solution propagates too slowly. Since a^* is a function of the wavenumber, the numerical approximation introduces dispersion, although the original PDE is nondispersive. As a result, a waveform consisting of many different wavenumber components eventually loses its original form.

Figure 3.5 shows the numerical phase speed for the schemes considered previously. The number of *points per wavelength* (PPW) by which a given wave is resolved is given by $2\pi/\kappa\Delta x$. The resolving efficiency of a scheme can be expressed in terms of the PPW required to produce errors below a specified level. For example, the second-order centered difference scheme requires 80 PPW to produce an error in phase speed of less than 0.1 %.

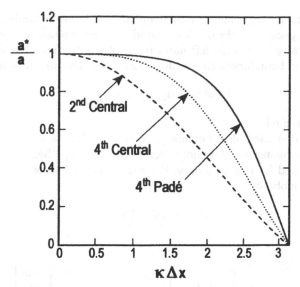

Fig. 3.5. Numerical phase speed for various schemes

The 5-point fourth-order centered scheme and the fourth-order Padé scheme require 15 and 10 PPW, respectively, to achieve the same error level.

For our example using second-order centered differences, the modified wavenumber is purely real, but in the general case it can include an imaginary component as well, as shown in (3.59). In that case, the error in the phase speed is determined from the real part of the modified wavenumber, while the imaginary part leads to an error in the amplitude of the solution, as can be seen by inspecting (3.61). Thus the antisymmetric portion of the spatial difference operator determines the error in speed and the symmetric portion the error in amplitude. This will be further discussed in Section 11.2.

3.6 Difference Operators at Boundaries

As discussed in Section 3.3.2, a matrix difference operator incorporates both the difference approximation in the interior of the domain and that at the boundaries. In this section, we consider the boundary operators needed for our model equations for convection and diffusion. In the case of periodic boundary conditions, no special boundary operators are required.

3.6.1 The Linear Convection Equation

Referring to Section 2.3, the boundary conditions for the linear convection equation can be either periodic or of inflow–outflow type. In the latter case, a Dirichlet boundary condition is given at the inflow boundary, while no condition is specified at the outflow boundary. Here we assume that the wave speed a is positive; thus the left-hand boundary is the inflow boundary, and the right-hand boundary is the outflow boundary. The vector of unknowns is

$$\vec{u} = [u_1,\ u_2,\ \ldots,\ u_M]^{\mathrm{T}} \tag{3.63}$$

and u_0 is specified.

Consider first the inflow boundary. It is clear that as long as the interior difference approximation does not extend beyond u_{j-1}, then no special treatment is required for this boundary. For example, with second-order centered differences we obtain

$$\delta_x \vec{u} = A\vec{u} + \left(\vec{bc}\right) \tag{3.64}$$

with

$$A = \frac{1}{2\Delta x} \begin{bmatrix} 0 & 1 & & \\ -1 & 0 & 1 & \\ & -1 & 0 & 1 \\ & & & \ddots \end{bmatrix}, \qquad \left(\vec{bc}\right) = \frac{1}{2\Delta x} \begin{bmatrix} -u_0 \\ 0 \\ 0 \\ \vdots \end{bmatrix}. \tag{3.65}$$

However, if we use the fourth-order interior operator given in (3.54), then the approximation at $j = 1$ requires a value of u_{j-2}, which is outside the domain. Hence, a different operator is required at $j = 1$ which extends only to $j - 1$, while having the appropriate order of accuracy. Such an operator, known as a *numerical boundary scheme*, can have an order of accuracy which is *one order lower* than that of the interior scheme, and the global accuracy will equal that of the interior scheme.[7] For example, with fourth-order centered differences, we can use the following third-order operator at $j = 1$:

$$(\delta_x u)_1 = \frac{1}{6\Delta x}(-2u_0 - 3u_1 + 6u_2 - u_3) , \tag{3.66}$$

which is easily derived using a Taylor table. The resulting difference operator has the form of (3.64) with

$$A = \frac{1}{12\Delta x} \begin{bmatrix} -6 & 12 & -2 & & \\ -8 & 0 & 8 & -1 & \\ 1 & -8 & 0 & 8 & -1 \\ & & & & \ddots \end{bmatrix}, \qquad \left(\vec{bc}\right) = \frac{1}{12\Delta x} \begin{bmatrix} -4u_0 \\ u_0 \\ 0 \\ \vdots \end{bmatrix}. \tag{3.67}$$

This approximation is globally fourth-order accurate.

---

[7] Proof of this theorem is beyond the scope of this book; the interested reader should consult the literature for further details.

At the outflow boundary, no boundary condition is specified. We must approximate $\partial u / \partial x$ at node M with no information about u_{M+1}. Thus the second-order centered-difference operator, which requires u_{j+1}, cannot be used at $j = M$. A backward-difference formula must be used. With a second-order interior operator, the following first-order backward formula can be used:

$$(\delta_x u)_M = \frac{1}{\Delta x}(u_M - u_{M-1}) . \tag{3.68}$$

This produces a difference operator with

$$A = \frac{1}{2\Delta x}\begin{bmatrix} 0 & 1 & & & & \\ -1 & 0 & 1 & & & \\ & -1 & 0 & 1 & & \\ & & & \ddots & & \\ & & & -1 & 0 & 1 \\ & & & & -2 & 2 \end{bmatrix}, \quad (\vec{bc}) = \frac{1}{2\Delta x}\begin{bmatrix} -u_0 \\ 0 \\ 0 \\ \vdots \\ 0 \end{bmatrix}. \tag{3.69}$$

In the case of a fourth-order centered interior operator the last two rows of A require modification.

Another approach to the development of boundary schemes is in terms of space extrapolation. The following formula allows u_{M+1} to be extrapolated from the interior data to arbitrary order on an equispaced grid:

$$(1 - E^{-1})^p u_{M+1} = 0 , \tag{3.70}$$

where E is the shift operator defined by $Eu_j = u_{j+1}$, and the order of the approximation is $p - 1$. For example, with $p = 2$ we obtain

$$(1 - 2E^{-1} + E^{-2})u_{M+1} = u_{M+1} - 2u_M + u_{M-1} = 0 , \tag{3.71}$$

which gives the following first-order approximation to u_{M+1}:

$$u_{M+1} = 2u_M - u_{M-1} . \tag{3.72}$$

Substituting this into the second-order centered-difference operator applied at node M gives

$$(\delta_x u)_M = \frac{1}{2\Delta x}(u_{M+1} - u_{M-1}) = \frac{1}{2\Delta x}(2u_M - u_{M-1} - u_{M-1})$$
$$= \frac{1}{\Delta x}(u_M - u_{M-1}) , \tag{3.73}$$

which is identical to (3.68).

3.6.2 The Diffusion Equation

In solving the diffusion equation, we must consider Dirichlet and Neumann boundary conditions. The treatment of a Dirichlet boundary condition proceeds along the same lines as the inflow boundary for the convection equation discussed above. With the second-order centered interior operator

$$(\delta_{xx}u)_j = \frac{1}{\Delta x^2}(u_{j+1} - 2u_j + u_{j-1}) \tag{3.74}$$

no modifications are required near boundaries, leading to the matrix difference operator given in (3.27).

For a Neumann boundary condition, we assume that $\partial u/\partial x$ is specified at $j = M + 1$, that is

$$\left(\frac{\partial u}{\partial x}\right)_{M+1} = \left(\frac{\partial u}{\partial x}\right)_b . \tag{3.75}$$

Thus we design an operator at node M which is in the following form:

$$(\delta_{xx}u)_M = \frac{1}{\Delta x^2}(au_{M-1} + bu_M) + \frac{c}{\Delta x}\left(\frac{\partial u}{\partial x}\right)_{M+1} , \tag{3.76}$$

where a, b, and c are constants which can easily be determined using a Taylor table, as shown in Table 3.4.

Solving for a, b, and c, we obtain the following first-order operator:

$$(\delta_{xx}u)_M = \frac{1}{3\Delta x^2}(2u_{M-1} - 2u_M) + \frac{2}{3\Delta x}\left(\frac{\partial u}{\partial x}\right)_{M+1} , \tag{3.77}$$

which produces the matrix difference operator given in (3.29).

We can also obtain the operator in (3.77) using the space extrapolation idea. Consider a second-order backward-difference approximation applied at node $M + 1$:

$$\left(\frac{\partial u}{\partial x}\right)_{M+1} = \frac{1}{2\Delta x}(u_{M-1} - 4u_M + 3u_{M+1}) + O(\Delta x^2) . \tag{3.78}$$

Solving for u_{M+1} gives

$$u_{M+1} = \frac{1}{3}\left[4u_M - u_{M-1} + 2\Delta x \left(\frac{\partial u}{\partial x}\right)_{M+1}\right] + O(\Delta x^3) . \tag{3.79}$$

Substituting this into the second-order centered difference operator for a second derivative applied at node M gives

Table 3.4. Taylor table for Neumann boundary condition for the diffusion equation

$$\left(\frac{\partial^2 u}{\partial x^2}\right)_j - \left[\frac{1}{\Delta x^2}(au_{j-1} + bu_j) + \frac{c}{\Delta x}\left(\frac{\partial u}{\partial x}\right)_{j+1}\right] = ?$$

	u_j	$\Delta x\left(\frac{\partial u}{\partial x}\right)_j$	$\Delta x^2\left(\frac{\partial^2 u}{\partial x^2}\right)_j$	$\Delta x^3\left(\frac{\partial^3 u}{\partial x^3}\right)_j$	$\Delta x^4\left(\frac{\partial^4 u}{\partial x^4}\right)_j$
$\Delta x^2\left(\frac{\partial^2 u}{\partial x^2}\right)_j$			1		
$-a\cdot u_{j-1}$	$-a$	$-a\cdot(-1)\cdot\frac{1}{1}$	$-a\cdot(-1)^2\cdot\frac{1}{2}$	$-a\cdot(-1)^3\cdot\frac{1}{6}$	$-a\cdot(-1)^4\cdot\frac{1}{24}$
$-b\cdot u_j$	$-b$				
$-\Delta x\cdot c\cdot\left(\frac{\partial u}{\partial x}\right)_{j+1}$		$-c$	$-c\cdot(1)\cdot\frac{1}{1}$	$-c\cdot(1)^2\cdot\frac{1}{2}$	$-c\cdot(1)^3\cdot\frac{1}{6}$

$$(\delta_{xx}u)_M = \frac{1}{\Delta x^2}(u_{M+1} - 2u_M + u_{M-1})$$

$$= \frac{1}{3\Delta x^2}\left[3u_{M-1} - 6u_M + 4u_M - u_{M-1} + 2\Delta x \left(\frac{\partial u}{\partial x}\right)_{M+1}\right]$$

$$= \frac{1}{3\Delta x^2}(2u_{M-1} - 2u_M) + \frac{2}{3\Delta x}\left(\frac{\partial u}{\partial x}\right)_{M+1}, \qquad (3.80)$$

which is identical to (3.77).

Exercises

3.1 Derive a third-order finite-difference approximation to a first derivative in the form

$$(\delta_x u)_j = \frac{1}{\Delta x}(au_{j-2} + bu_{j-1} + cu_j + du_{j+1}) \,.$$

Find the leading error term.

3.2 Derive a finite-difference approximation to a first derivative in the form

$$a(\delta_x u)_{j-1} + (\delta_x u)_j = \frac{1}{\Delta x}(bu_{j-1} + cu_j + du_{j+1}) \,.$$

Find the leading error term.

3.3 Using a 4 (interior) point mesh, write out the 4×4 matrices and the boundary-condition vector formed by using the scheme derived in Exercise 3.2 when both u and $\partial u/\partial x$ are given at $j = 0$ and u is given at $j = 5$.

3.4 Repeat Exercise 3.2 with $d = 0$.

3.5 Derive a finite-difference approximation to a third derivative in the form

$$(\delta_{xxx}u)_j = \frac{1}{\Delta x^3}(au_{j-2} + bu_{j-1} + cu_j + du_{j+1} + eu_{j+2}) \,.$$

Find the leading error term.

3.6 Derive a compact (or Padé) finite-difference approximation to a second derivative in the form

$$d(\delta_{xx}u)_{j-1} + (\delta_{xx}u)_j + e(\delta_{xx}u)_{j+1} = \frac{1}{\Delta x^2}(au_{j-1} + bu_j + cu_{j+1}) \,.$$

Find the leading error term.

3.7 Find the modified wavenumber for the operator derived in Exercise 3.1. Plot the real and imaginary parts of $\kappa^*\Delta x$ vs. $\kappa\Delta x$ for $0 \le \kappa\Delta x \le \pi$. Compare the real part with that obtained from the fourth-order centered operator (3.54).

3.8 Application of the second-derivative operator to the function $e^{i\kappa x}$ gives

$$\frac{\partial^2 e^{i\kappa x}}{\partial x^2} = -\kappa^2 e^{i\kappa x} \ .$$

Application of a difference operator for the second derivative gives

$$\left(\delta_{xx} e^{i\kappa j \Delta x}\right)_j = -\kappa^{*2} e^{i\kappa x}$$

thus defining the modified wavenumber κ^* for a second derivative approximation. Find the modified wavenumber for the second-order centered difference operator for a second derivative, the noncompact fourth-order operator (3.55), and the compact fourth-order operator derived in Exercise 3.6. Plot $(\kappa^* \Delta x)^2$ vs. $\kappa \Delta x$ for $0 \le \kappa \Delta x \le \pi$.

3.9 Find the grid-points-per-wavelength (PPW) requirement to achieve a phase speed error less than 0.1 percent for sixth-order noncompact and compact centered approximations to a first derivative.

3.10 Consider the following one-sided differencing schemes, which are first-, second-, and third-order, respectively:

$$(\delta_x u)_j = (u_j - u_{j-1})/\Delta x$$
$$(\delta_x u)_j = (3u_j - 4u_{j-1} + u_{j-2})/(2\Delta x)$$
$$(\delta_x u)_j = (11u_j - 18u_{j-1} + 9u_{j-2} - 2u_{j-3})/(6\Delta x) \ .$$

Find the modified wavenumber for each of these schemes. Plot the real and imaginary parts of $\kappa^* \Delta x$ vs. $\kappa \Delta x$ for $0 \le \kappa \Delta x \le \pi$. Derive the two leading terms in the truncation error for each scheme.

4. The Semi-Discrete Approach

One strategy for obtaining finite-difference approximations to a PDE is to start by differencing the space derivatives only, without approximating the time derivative. In the following chapters, we proceed with an analysis making considerable use of this concept, which we refer to as the *semi-discrete* approach. Differencing the space derivatives converts the basic PDE into a set of coupled ODE's. In the most general notation, these ODE's would be expressed in the form

$$\frac{d\vec{u}}{dt} = \vec{F}(\vec{u}, t) , \qquad (4.1)$$

which includes all manner of nonlinear and time-dependent possibilities. On occasion, we use this form, but the rest of this chapter is devoted to a more specialized matrix notation described below.

Another strategy for constructing a finite-difference approximation to a PDE is to approximate all the partial derivatives at once. This generally leads to a point difference operator (see Section 3.3.1) which, in turn, can be used for the time advance of the solution at any given point in the mesh. As an example let us consider the model equation for diffusion

$$\frac{\partial u}{\partial t} - \nu \frac{\partial^2 u}{\partial x^2} .$$

Using three-point central-differencing schemes for both the time and space derivatives, we find

$$\frac{u_j^{(n+1)} - u_j^{(n-1)}}{2h} = \nu \left(\frac{u_{j+1}^{(n)} - 2u_j^{(n)} + u_{j-1}^{(n)}}{\Delta x^2} \right)$$

or

$$u_j^{(n+1)} = u_j^{(n-1)} + \frac{2h\nu}{\Delta x^2} \left(u_{j+1}^{(n)} - 2u_j^{(n)} + u_{j-1}^{(n)} \right) . \qquad (4.2)$$

Clearly (4.2) is a difference equation which can be used at the space point j to advance the value of u from the previous time levels n and $n - 1$ to the level $n + 1$. It is a full discretization of the PDE. Note, however, that

the spatial and temporal discretizations are separable. Thus, this method
has an intermediate semi-discrete form and can be analyzed by the methods
discussed in the next few chapters.

Another possibility is to replace the value of $u_j^{(n)}$ in the right-hand side of
(4.2) with the time average of u at that point, namely $(u_j^{(n+1)} + u_j^{(n-1)})/2$.
This results in the formula

$$u_j^{(n+1)} = u_j^{(n-1)} + \frac{2h\nu}{\Delta x^2}\left[u_{j+1}^{(n)} - 2\left(\frac{u_j^{(n+1)} + u_j^{(n-1)}}{2}\right) + u_{j-1}^{(n)}\right], \quad (4.3)$$

which can be solved for $u^{(n+1)}$ and time advanced at the point j. In this
case, the spatial and temporal discretizations are not separable, and no semi-
discrete form exists.

Equation (4.2) is sometimes called Richardson's method of overlapping
steps and (4.3) is referred to as the DuFort–Frankel method. As we shall
see later on, there are subtle points to be made about using these methods
to find a numerical solution to the diffusion equation. There are a number
of issues concerning the accuracy, stability, and convergence of (4.2) and
(4.3) which we cannot comment on until we develop a framework for such
investigations. We introduce these methods here only to distinguish between
methods in which the temporal and spatial terms are discretized separately
and those for which no such separation is possible. For the time being, we
shall separate the space difference approximations from the time differencing.
In this approach, we reduce the governing PDE's to ODE's by discretizing
the spatial terms and use the well-developed theory of ODE solutions to aid
us in the development of an analysis of accuracy and stability.

4.1 Reduction of PDE's to ODE's

4.1.1 The Model ODE's

First let us consider the model PDE's for diffusion and biconvection described
in Chapter 2. In these simple cases, we can approximate the space derivatives
with difference operators and express the resulting ODE's with a matrix
formulation. This is a simple and natural formulation when the ODE's are
linear.

Model ODE for Diffusion. For example, using the three-point central-
differencing scheme to represent the second derivative in the scalar PDE
governing diffusion leads to the following ODE diffusion model

$$\frac{d\vec{u}}{dt} = \frac{\nu}{\Delta x^2}B(1,-2,1)\vec{u} + (\vec{bc}) \quad (4.4)$$

with Dirichlet boundary conditions folded into the $(\vec{bc})$ vector.

Model ODE for Biconvection. The term biconvection was introduced in Section 2.3. It is used for the scalar convection model when the boundary conditions are periodic. In this case, the 3-point central-differencing approximation produces the ODE model given by

$$\frac{\mathrm{d}\vec{u}}{\mathrm{d}t} = -\frac{a}{2\Delta x}B_{\mathrm{p}}(-1, 0, 1)\vec{u} , \tag{4.5}$$

where the boundary condition vector is absent because the flow is periodic.

Equations (4.4) and (4.5) are the model ODE's for diffusion and biconvection of a scalar in one dimension. They are linear with coefficient matrices which are independent of x and t.

4.1.2 The Generic Matrix Form

The generic matrix form of a semi-discrete approximation is expressed by the equation

$$\boxed{\frac{\mathrm{d}\vec{u}}{\mathrm{d}t} = A\vec{u} - \vec{f}(t) .} \tag{4.6}$$

Note that the elements in the matrix A depend upon both the PDE and the type of differencing scheme chosen for the space terms. The vector $\vec{f}(t)$ is usually determined by the boundary conditions and possibly source terms. In general, even the Euler and Navier–Stokes equations can be expressed in the form of (4.6). In such cases the equations are nonlinear, that is, the elements of A depend on the solution $\vec{u}$ and are usually derived by finding the Jacobian of a flux vector. Although the equations are nonlinear, the linear analysis presented in this book leads to diagnostics that are surprisingly accurate when used to evaluate many aspects of numerical methods as they apply to the Euler and Navier–Stokes equations.

4.2 Exact Solutions of Linear ODE's

In order to advance (4.1) in time, the system of ODE's must be integrated using a time-marching method. In order to analyze time-marching methods, we will make use of exact solutions of coupled systems of ODE's, which exist under certain conditions. The ODE's represented by (4.1) are said to be *linear* if F is linearly dependent on u (i.e. if $\partial F/\partial u = A$, where A is independent of u). As we have already pointed out, when the ODE's are linear they can be expressed in a matrix notation as (4.6) in which the coefficient matrix, A, is independent of u. If A *does* depend explicitly on t, the general solution *cannot* be written, whereas, if A *does not* depend explicitly on t, the general solution to (4.6) *can* be written. This holds regardless of whether or not the forcing function, $\vec{f}$, depends explicitly on t.

As we shall soon see, the exact solution of (4.6) can be written in terms of the eigenvalues and eigenvectors of A. This will lead us to a representative scalar equation for use in analyzing time-marching methods. These ideas are developed in the following sections.

4.2.1 Eigensystems of Semi-discrete Linear Forms

Complete Systems. An $M \times M$ matrix is represented by a *complete eigensystem* if it has a complete set of linearly independent eigenvectors (see Appendix A). An eigenvector, $\vec{x}_m$, and its corresponding eigenvalue, λ_m, have the property that

$$A\vec{x}_m = \lambda_m \vec{x}_m$$

or

$$[A - \lambda_m I]\vec{x}_m = 0 . \qquad (4.7)$$

The eigenvalues are the roots of the equation

$$\det[A - \lambda I] = 0 .$$

We form the right-hand eigenvector matrix of a complete system by filling its columns with the eigenvectors $\vec{x}_m$:

$$X = [\vec{x}_1 , \ \vec{x}_2 \ \dots \ , \vec{x}_M] .$$

The inverse is the left-hand eigenvector matrix, and together they have the property that

$$X^{-1}AX = \Lambda , \qquad (4.8)$$

where Λ is a diagonal matrix whose elements are the eigenvalues of A.

Defective Systems. If an $M \times M$ matrix does *not* have a complete set of linearly independent eigenvectors, it cannot be transformed to a diagonal matrix of scalars, and it is said to be defective. It can, however, be transformed to a diagonal set of *blocks*, some of which may be scalars (see Appendix A). In general, there exists some S which transforms any matrix A such that

$$S^{-1}AS = J ,$$

where

$$J = \begin{bmatrix} J_1 & & & & \\ & J_2 & & & \\ & & \ddots & & \\ & & & J_m & \\ & & & & \ddots \end{bmatrix}$$

and

$$J_m^{(n)} = \begin{bmatrix} \lambda_m & 1 & & \\ & \lambda_m & \ddots & \\ & & \ddots & 1 \\ & & & \lambda_m \end{bmatrix} \begin{matrix} 1 \\ \vdots \\ \vdots \\ n \end{matrix}$$

The matrix J is said to be in *Jordan canonical form*, and an eigenvalue with multiplicity n within a Jordan block is said to be a defective eigenvalue. Defective systems play a role in numerical stability analysis.

4.2.2 Single ODE's of First and Second Order

First-Order Equations. The simplest nonhomogeneous ODE of interest is given by the single, first-order equation

$$\frac{du}{dt} = \lambda u + a e^{\mu t} , \tag{4.9}$$

where λ, a, and μ are scalars, all of which can be complex numbers. The equation is linear because λ does not depend on u, and has a general solution because λ does not depend on t. It has a steady-state solution if the right-hand side is independent of t, i.e. if $\mu = 0$, and is homogeneous if the forcing function is zero, i.e. if $a = 0$. Although it is quite simple, the numerical analysis of (4.9) displays many of the fundamental properties and issues involved in the construction and study of most popular time-marching methods. This theme will be developed as we proceed.

The exact solution of (4.9) is, for $\mu \neq \lambda$,

$$u(t) = c_1 e^{\lambda t} + \frac{a e^{\mu t}}{\mu - \lambda} ,$$

where c_1 is a constant determined by the initial conditions. In terms of the initial value of u, it can be written

$$u(t) = u(0) e^{\lambda t} + a \frac{e^{\mu t} - e^{\lambda t}}{\mu - \lambda} .$$

The interesting question can arise: What happens to the solution of (4.9) when $\mu = \lambda$? This is easily found by setting $\mu = \lambda + \epsilon$, solving, and then taking the limit as $\epsilon \to 0$. Using this limiting device, we find that the solution to

$$\frac{du}{dt} = \lambda u + a e^{\lambda t} \tag{4.10}$$

is given by

$$u(t) = [u(0) + at] e^{\lambda t} .$$

As we shall soon see, this solution is required for the analysis of defective systems.

Second-Order Equations. The homogeneous form of a second-order equation is given by

$$\frac{\mathrm{d}^2 u}{\mathrm{d}t^2} + a_1 \frac{\mathrm{d}u}{\mathrm{d}t} + a_0 u = 0 . \tag{4.11}$$

where a_1 and a_0 are complex constants. Now we introduce the differential operator D such that

$$\boxed{D \equiv \frac{\mathrm{d}}{\mathrm{d}t}}$$

and factor $u(t)$ out of (4.11), giving

$$(D^2 + a_1 D + a_0)\, u(t) = 0 .$$

The polynomial in D is referred to as a *characteristic polynomial* and designated $P(D)$. Characteristic polynomials are fundamental to the analysis of both ODE's and ordinary difference equations (OΔE's), since the roots of these polynomials determine the solutions of the equations. For ODE's, we often label these roots in order of increasing modulus as $\lambda_1, \lambda_2, \cdots, \lambda_m,$ $\cdots, \lambda_M$. They are found by solving the equation $P(\lambda) = 0$. In our simple example, there would be two roots, λ_1 and λ_2, determined from

$$P(\lambda) = \lambda^2 + a_1 \lambda + a_0 = 0 \tag{4.12}$$

and the solution to (4.11) is given by

$$u(t) = c_1 e^{\lambda_1 t} + c_2 e^{\lambda_2 t} , \tag{4.13}$$

where c_1 and c_2 are constants determined from initial conditions. The proof of this is simple and is found by substituting (4.13) into (4.11). One finds the result

$$c_1 e^{\lambda_1 t}(\lambda_1^2 + a_1 \lambda_1 + a_0) + c_2 e^{\lambda_2 t}(\lambda_2^2 + a_1 \lambda_2 + a_0) ,$$

which is identically zero for all c_1, c_2, and t if and only if the λ's satisfy (4.12).

4.2.3 Coupled First-Order ODE's

A Complete System. A set of coupled, first-order, homogeneous equations is given by

$$u_1' = a_{11} u_1 + a_{12} u_2 \tag{4.14}$$
$$u_2' = a_{21} u_1 + a_{22} u_2 ,$$

which can be written as

$$\vec{u}' = A\vec{u} \quad , \quad \vec{u} = [u_1, u_2]^{\mathrm{T}} \quad , \quad A = (a_{ij}) = \begin{bmatrix} a_{11} & a_{12} \\ a_{21} & a_{22} \end{bmatrix} .$$

Consider the possibility that a solution is represented by

$$u_1 = c_1 \, x_{11} e^{\lambda_1 t} + c_2 \, x_{12} e^{\lambda_2 t}$$ (4.15)
$$u_2 = c_1 \, x_{21} e^{\lambda_1 t} + c_2 \, x_{22} e^{\lambda_2 t} .$$

By substitution, these are indeed solutions to (4.14) if and only if

$$\begin{bmatrix} a_{11} & a_{12} \\ a_{21} & a_{22} \end{bmatrix} \begin{bmatrix} x_{11} \\ x_{21} \end{bmatrix} = \lambda_1 \begin{bmatrix} x_{11} \\ x_{21} \end{bmatrix}, \quad \begin{bmatrix} a_{11} & a_{12} \\ a_{21} & a_{22} \end{bmatrix} \begin{bmatrix} x_{12} \\ x_{22} \end{bmatrix} = \lambda_2 \begin{bmatrix} x_{12} \\ x_{22} \end{bmatrix} .$$ (4.16)

Notice that a higher-order equation can be reduced to a coupled set of first-order equations by introducing a new set of dependent variables. Thus, by setting

$$u_1 = u' , \quad u_2 = u ,$$

we find (4.11) can be written as

$$u_1' = -a_1 u_1 - a_0 u_2$$ (4.17)
$$u_2' = u_1 ,$$

which is a subset of (4.14).

A Derogatory System. Equation (4.15) is still a solution to (4.14) if $\lambda_1 = \lambda_2 = \lambda$, provided two linearly independent vectors exist to satisfy (4.16) with $A = \Lambda$. In this case

$$\begin{bmatrix} \lambda & 0 \\ 0 & \lambda \end{bmatrix} \begin{bmatrix} 1 \\ 0 \end{bmatrix} = \lambda \begin{bmatrix} 1 \\ 0 \end{bmatrix} \quad \text{and} \quad \begin{bmatrix} \lambda & 0 \\ 0 & \lambda \end{bmatrix} \begin{bmatrix} 0 \\ 1 \end{bmatrix} = \lambda \begin{bmatrix} 0 \\ 1 \end{bmatrix}$$

provide such a solution. This is the case where A has a complete set of eigenvectors and is not defective.

A Defective System. If A is defective, then it can be represented by the *Jordan canonical form*

$$\begin{bmatrix} u_1' \\ u_2' \end{bmatrix} = \begin{bmatrix} \lambda & 0 \\ 1 & \lambda \end{bmatrix} \begin{bmatrix} u_1 \\ u_2 \end{bmatrix} ,$$ (4.18)

whose solution is not obvious. However, in this case, one can solve the top equation first, giving $u_1(t) = u_1(0)e^{\lambda t}$. Then, substituting this result into the second equation, one finds

$$\frac{du_2}{dt} = \lambda u_2 + u_1(0)e^{\lambda t} ,$$

which is identical in form to (4.10) and has the solution

$$u_2(t) = [u_2(0) + u_1(0)t]e^{\lambda t} .$$

From this the reader should be able to verify that

$$u_3(t) = (a + bt + ct^2)e^{\lambda t}$$

is a solution to

$$\begin{bmatrix} u_1' \\ u_2' \\ u_3' \end{bmatrix} = \begin{bmatrix} \lambda & & \\ 1 & \lambda & \\ & 1 & \lambda \end{bmatrix} \begin{bmatrix} u_1 \\ u_2 \\ u_3 \end{bmatrix}$$

if

$$a = u_3(0) \quad , \quad b = u_2(0) \quad , \quad c = \frac{1}{2}u_1(0) \ . \tag{4.19}$$

The general solution to such defective systems is left as an exercise.

4.2.4 General Solution of Coupled ODE's with Complete Eigensystems

Let us consider a set of coupled, nonhomogeneous, linear, first-order ODE's with constant coefficients which might have been derived by space differencing a set of PDE's. Represent them by the equation

$$\frac{d\vec{u}}{dt} = A\vec{u} - \vec{f}(t) \ . \tag{4.20}$$

Our assumption is that the $M \times M$ matrix A has a complete eigensystem[1] and can be transformed by the left and right eigenvector matrices, X^{-1} and X, to a diagonal matrix Λ having diagonal elements which are the eigenvalues of A (see Section 4.2.1). Now let us multiply (4.20) from the left by X^{-1} and insert the identity combination $XX^{-1} = I$ between A and $\vec{u}$. There results

$$X^{-1}\frac{d\vec{u}}{dt} = X^{-1}AX \cdot X^{-1}\vec{u} - X^{-1}\vec{f}(t) \ . \tag{4.21}$$

Since A is independent of both $\vec{u}$ and t, the elements in X^{-1} and X are also independent of both $\vec{u}$ and t, and (4.21) can be modified to

$$\frac{d}{dt}X^{-1}\vec{u} = \Lambda X^{-1}\vec{u} - X^{-1}\vec{f}(t) \ .$$

Finally, by introducing the new variables $\vec{w}$ and $\vec{g}$ such that

$$\vec{w} = X^{-1}\vec{u} \quad , \quad \vec{g}(t) = X^{-1}\vec{f}(t) \ , \tag{4.22}$$

we reduce (4.20) to a new algebraic form

---

[1] In the following, we exclude defective systems, not because they cannot be analyzed (the example at the conclusion of the previous section proves otherwise), but because they are only of limited interest in the general development of our theory.

$$\frac{d\vec{w}}{dt} = \Lambda\vec{w} - \vec{g}(t) \ . \tag{4.23}$$

It is important at this point to review the results of the previous paragraph. Notice that (4.20) and (4.23) are expressing exactly the same equality. The only difference between them was brought about by algebraic manipulations which regrouped the variables. However, this regrouping is crucial for the solution process because the equations represented by (4.23) are no longer coupled. They can be written line by line as a set of independent, single, first-order equations, thus

$$w'_1 = \lambda_1 w_1 - g_1(t)$$
$$\vdots$$
$$w'_m = \lambda_m w_m - g_m(t) \tag{4.24}$$
$$\vdots$$
$$w'_M = \lambda_M w_M - g_M(t) \ .$$

For any given set of $g_m(t)$ each of these equations can be solved separately and then recoupled, using the inverse of the relations given in (4.22):

$$\vec{u}(t) = X\vec{w}(t)$$
$$= \sum_{m=1}^{M} w_m(t)\vec{x}_m \ , \tag{4.25}$$

where $\vec{x}_m$ is the mth column of X, i.e. the eigenvector corresponding to λ_m.

We next focus on the very important subset of (4.20) when neither A nor $\vec{f}$ has any explicit dependence on t. In such a case, the g_m in (4.23) and (4.24) are also time invariant and the solution to any line in (4.24) is

$$w_m(t) = c_m e^{\lambda_m t} + \frac{1}{\lambda_m} g_m \ ,$$

where the c_m are constants that depend on the initial conditions. Transforming back to the u-system gives

$$\vec{u}(t) = X\vec{w}(t)$$
$$= \sum_{m=1}^{M} w_m(t)\vec{x}_m$$
$$= \sum_{m=1}^{M} c_m e^{\lambda_m t} \vec{x}_m + \sum_{m=1}^{M} \frac{1}{\lambda_m} g_m \vec{x}_m$$
$$= \sum_{m=1}^{M} c_m e^{\lambda_m t} \vec{x}_m + X\Lambda^{-1}X^{-1}\vec{f}$$

$$= \sum_{m=1}^{M} c_m e^{\lambda_m t} \, \vec{x}_m + \underbrace{A^{-1}\vec{f}}. \tag{4.26}$$

$$\underbrace{\phantom{\sum_{m=1}^{M} c_m e^{\lambda_m t} \, \vec{x}_m}}_{\text{transient}} \quad \underbrace{\phantom{A^{-1}\vec{f}}}_{\text{steady-state}}$$

Note that the steady-state solution is $A^{-1}\vec{f}$, as might be expected.

The first group of terms on the right side of this equation is referred to classically as the *complementary solution* or the solution of the homogeneous equations. The second group is referred to classically as the *particular solution* or the particular integral. In our application to fluid dynamics, it is more instructive to refer to these groups as the *transient* and *steady-state* solutions, respectively. An alternative, but entirely equivalent, form of the solution is

$$\vec{u}(t) = c_1 e^{\lambda_1 t} \, \vec{x}_1 + \cdots + c_m e^{\lambda_m t} \, \vec{x}_m + \cdots + c_M e^{\lambda_M t} \, \vec{x}_M + A^{-1}\vec{f}. \tag{4.27}$$

4.3 Real Space and Eigenspace

4.3.1 Definition

Following the semi-discrete approach discussed in Section 4.1, we reduce the partial differential equations to a set of ordinary differential equations represented by the generic form

$$\frac{d\vec{u}}{dt} = A\vec{u} - \vec{f}. \tag{4.28}$$

The dependent variable $\vec{u}$ represents some physical quantity or quantities which relate to the problem of interest. For the model problems on which we are focusing most of our attention, the elements of A are independent of both u and t. This permits us to say a great deal about such problems and serves as the basis for this section. In particular, we can develop some very important and fundamental concepts that underly the *global* properties of the numerical solutions to the model problems. How these relate to the numerical solutions of more practical problems depends upon the problem and, to a much greater extent, on the cleverness of the relator.

We begin by developing the concept of "spaces". That is, we identify different mathematical reference frames (spaces) and view our solutions from within each. In this way, we get different perspectives of the same solution, and this can add significantly to our understanding.

The most natural reference frame is the physical one. We say:

> If a solution is expressed in terms of $\vec{u}$, it is said
> to be in *real space*.

There is, however, another very useful frame. We saw in Section 4.2 that pre-
and post-multiplication of A by the appropriate similarity matrices transforms A into a diagonal matrix, composed, in the most general case, of Jordan
blocks or, in the simplest nondefective case, of scalars. Following Section 4.2
and, for simplicity, using only complete systems for our examples, we found
that (4.28) had the alternative form

$$\frac{d\vec{w}}{dt} = \Lambda\vec{w} - \vec{g},$$

which is an uncoupled set of first-order ODE's that can be solved indepen-
dently for the dependent variable vector $\vec{w}$. We say:

> If a solution is expressed in terms of $\vec{w}$, it is said to
> be in *eigenspace* (often referred to as wave space).

The relations that transfer from one space to the other are:

$$\vec{w} = X^{-1}\vec{u} \qquad\qquad \vec{u} = X\vec{w}$$
$$\vec{g} = X^{-1}\vec{f} \qquad\qquad \vec{f} = X\vec{g}$$

The elements of $\vec{u}$ relate directly to the local physics of the problem. However,
the elements of $\vec{w}$ are linear combinations of all of the elements of $\vec{u}$, and
individually they have no direct local physical interpretation.

When the forcing function $\vec{f}$ is independent of t, the solutions in the two
spaces are represented by

$$\vec{u}(t) = \sum c_m e^{\lambda_m t}\, \vec{x}_m + X\Lambda^{-1}X^{-1}\vec{f}$$

and

$$w_m(t) = c_m e^{\lambda_m t} + \frac{1}{\lambda_m}g_m, \quad m = 1, 2, \cdots, M$$

for real space and eigenspace, respectively. At this point we make two observations:

(1) The transient portion of the solution in real space consists of a linear
combination of contributions from each eigenvector, and

(2) The transient portion of the solution in eigenspace provides the weight-
ing of each eigenvector component of the transient solution in real
space.

4.3.2 Eigenvalue Spectrums for Model ODE's

It is instructive to consider the eigenvalue spectrums of the ODE's formulated
by central differencing the model equations for diffusion and biconvection.

These model ODE's are presented in Section 4.1. Equations for the eigenvalues of the simple tridiagonals, $B(M : a, b, c)$ and $B_p(M : a, b, c)$, are given in Appendix B. From Appendix B.1, we find for the model diffusion equation with Dirichlet boundary conditions

$$\lambda_m = \frac{\nu}{\Delta x^2}\left[-2 + 2\cos\left(\frac{m\pi}{M+1}\right)\right]$$

$$= \frac{-4\nu}{\Delta x^2}\sin^2\left(\frac{m\pi}{2(M+1)}\right), \quad m = 1, 2, \cdots, M \tag{4.29}$$

and, from Appendix B.4, for the model biconvection equation

$$\lambda_m = \frac{-ia}{\Delta x}\sin\left(\frac{2m\pi}{M}\right), \quad m = 0, 1, \cdots, M-1$$

$$= -i\kappa_m^* a , \tag{4.30}$$

where

$$\kappa_m^* = \frac{\sin\kappa_m\Delta x}{\Delta x}, \quad m = 0, 1, \cdots, M-1 \tag{4.31}$$

is the modified wavenumber from Section 3.5, $\kappa_m = m$, and $\Delta x = 2\pi/M$. Notice that the diffusion eigenvalues are real and negative while those representing periodic convection are all pure imaginary. The interpretation of this result plays a very important role later in our stability analysis.

4.3.3 Eigenvectors of the Model Equations

Next we consider the eigenvectors of the two model equations. These follow as special cases from the results given in Appendix B.

The Diffusion Model. Consider (4.4), the model ODE's for diffusion. First, to help visualize the matrix structure, we present results for a simple 4-point mesh, and then we give the general case. The right-hand eigenvector matrix X is given by

$$\begin{bmatrix} \sin(x_1) & \sin(2x_1) & \sin(3x_1) & \sin(4x_1) \\ \sin(x_2) & \sin(2x_2) & \sin(3x_2) & \sin(4x_2) \\ \sin(x_3) & \sin(2x_3) & \sin(3x_3) & \sin(4x_3) \\ \sin(x_4) & \sin(2x_4) & \sin(3x_4) & \sin(4x_4) \end{bmatrix}.$$

The columns of the matrix are the right eigenvectors. Recall that $x_j = j\Delta x = j\pi/(M+1)$, so in general the relation $\vec{u} = X\vec{w}$ can be written as

$$u_j = \sum_{m=1}^{M} w_m \sin mx_j, \quad j = 1, 2, \cdots, M . \tag{4.32}$$

For the inverse, or left-hand eigenvector matrix X^{-1}, we find

$$\begin{bmatrix} \sin(x_1) & \sin(x_2) & \sin(x_3) & \sin(x_4) \\ \sin(2x_1) & \sin(2x_2) & \sin(2x_3) & \sin(2x_4) \\ \sin(3x_1) & \sin(3x_2) & \sin(3x_3) & \sin(3x_4) \\ \sin(4x_1) & \sin(4x_2) & \sin(4x_3) & \sin(4x_4) \end{bmatrix} .$$

The rows of the matrix are the left eigenvectors. In general $\vec{w} = X^{-1}\vec{u}$ gives

$$w_m = \sum_{j=1}^{M} u_j \sin m x_j, \quad m = 1, 2, \cdots, M . \tag{4.33}$$

In the field of harmonic analysis, (4.33) represents a sine transform of the function $u(x)$ for an M-point sample between the boundaries $x = 0$ and $x = \pi$ with the condition $u(0) = u(\pi) = 0$. Similarly, (4.32) represents the sine synthesis that companions the sine transform given by (4.33). In summary:

For the model diffusion equation:

$\vec{w} = X^{-1}\vec{u}$ is a sine transform from real space to (sine) wave space.

$\vec{u} = X\vec{w}$ is a sine synthesis from wave space back to real space.

The Biconvection Model. Next consider the model ODE's for periodic convection, (4.5). The coefficient matrices for these ODE's are always circulant. For our model ODE, the right-hand eigenvectors are given by

$$\vec{x}_m = \mathrm{e}^{\mathrm{i}\, j\, (2\pi m/M)}, \quad \begin{array}{l} j = 0, 1, \cdots, M - 1 \\ m = 0, 1, \cdots, M - 1 \end{array} .$$

With $x_j = j \cdot \Delta x = j \cdot 2\pi/M$, we can write $\vec{u} = X\vec{w}$ as

$$u_j = \sum_{m=0}^{M-1} w_m \mathrm{e}^{\mathrm{i} m x_j}, \quad j = 0, 1, \cdots, M - 1 . \tag{4.34}$$

For a 4-point periodic mesh, we find the following left-hand eigenvector matrix from Appendix B.4:

$$\begin{bmatrix} w_1 \\ w_2 \\ w_3 \\ w_4 \end{bmatrix} = \frac{1}{4} \begin{bmatrix} 1 & 1 & 1 & 1 \\ 1 & \mathrm{e}^{-2\mathrm{i}\pi/4} & \mathrm{e}^{-4\mathrm{i}\pi/4} & \mathrm{e}^{-6\mathrm{i}\pi/4} \\ 1 & \mathrm{e}^{-4\mathrm{i}\pi/4} & \mathrm{e}^{-8\mathrm{i}\pi/4} & \mathrm{e}^{-12\mathrm{i}\pi/4} \\ 1 & \mathrm{e}^{-6\mathrm{i}\pi/4} & \mathrm{e}^{-12\mathrm{i}\pi/4} & \mathrm{e}^{-18\mathrm{i}\pi/4} \end{bmatrix} \begin{bmatrix} u_1 \\ u_2 \\ u_3 \\ u_4 \end{bmatrix} = X^{-1}\vec{u} .$$

In general

$$w_m = \frac{1}{M} \sum_{j=0}^{M-1} u_j e^{-imx_j}, \quad m = 0, 1, \cdots, M - 1 .$$

This equation is identical to a discrete Fourier transform of the periodic dependent variable $\vec{u}$ using an M-point sample between and including $x = 0$ and $x = 2\pi - \Delta x$.

For circulant matrices, it is straightforward to establish the fact that the relation $\vec{u} = X\vec{w}$ represents the Fourier synthesis of the variable $\vec{w}$ back to $\vec{u}$. In summary:

For any circulant system:

$\vec{w} = X^{-1}\vec{u}$ is a complex Fourier transform from real space to wave space.

$\vec{u} = X\vec{w}$ is a complex Fourier synthesis from wave space back to real space.

4.3.4 Solutions of the Model ODE's

We can now combine the results of the previous sections to write the solutions of our model ODE's.

The Diffusion Equation. For the diffusion equation, (4.27) becomes

$$u_j(t) = \sum_{m=1}^{M} c_m e^{\lambda_m t} \sin mx_j + (A^{-1}f)_j, \qquad j = 1, 2, \cdots, M , \quad (4.35)$$

where

$$\lambda_m = \frac{-4\nu}{\Delta x^2} \sin^2 \left(\frac{m\pi}{2(M+1)} \right) . \qquad (4.36)$$

With the modified wavenumber defined as

$$\kappa_m^* = \frac{2}{\Delta x} \sin \left(\frac{\kappa_m \Delta x}{2} \right) \qquad (4.37)$$

and using $\kappa_m = m$, we can write the ODE solution as

$$u_j(t) = \sum_{m=1}^{M} c_m e^{-\nu \kappa_m^{*2} t} \sin \kappa_m x_j + (A^{-1}f)_j, \qquad j = 1, 2, \cdots, M . \quad (4.38)$$

This can be compared with the exact solution to the PDE, (2.37), evaluated at the nodes of the grid:

$$u_j(t) = \sum_{m=1}^{M} c_m e^{-\nu \kappa_m{}^2 t} \sin \kappa_m x_j + h(x_j), \qquad j = 1, 2, \cdots, M . \quad (4.39)$$

We see that the solutions are identical except for the steady solution and the modified wavenumber in the transient term. The modified wavenumber is an approximation to the actual wavenumber. The difference between the modified wavenumber and the actual wavenumber depends on the differencing scheme and the grid resolution. This difference causes the various modes (or eigenvector components) to decay at rates which differ from the exact solution. With conventional differencing schemes, low wavenumber modes are accurately represented, while high wavenumber modes (if they have significant amplitudes) can have large errors.

The Convection Equation. For the biconvection equation, we obtain

$$u_j(t) = \sum_{m=0}^{M-1} c_m e^{\lambda_m t} e^{i\kappa_m x_j}, \qquad j = 0, 1, \cdots, M - 1 , \quad (4.40)$$

where

$$\lambda_m = -i\kappa_m^* a \quad (4.41)$$

with the modified wavenumber defined in (4.31). We can write this ODE solution as

$$u_j(t) = \sum_{m=0}^{M-1} c_m e^{-i\kappa_m^* a t} e^{i\kappa_m x_j}, \qquad j = 0, 1, \cdots, M - 1 \quad (4.42)$$

and compare it to the exact solution of the PDE, (2.26), evaluated at the nodes of the grid:

$$u_j(t) = \sum_{m=0}^{M-1} f_m(0) e^{-i\kappa_m a t} e^{i\kappa_m x_j}, \qquad j = 0, 1, \cdots, M - 1 . \quad (4.43)$$

Once again the difference appears through the modified wavenumber contained in λ_m. As discussed in Section 3.5, this leads to an error in the speed with which various modes are convected, since κ^* is real. Since the error in the phase speed depends on the wavenumber, while the actual phase speed is independent of the wavenumber, the result is erroneous numerical dispersion. In the case of non-centered differencing, discussed in Chapter 11, the modified wavenumber is complex. The form of (4.42) shows that the imaginary portion of the modified wavenumber produces nonphysical decay or growth in the numerical solution.

4.4 The Representative Equation

In Section 4.3, we pointed out that (4.20) and (4.23) express identical results but in terms of different groupings of the dependent variables, which are related by algebraic manipulation. This leads to the following important concept:

> The numerical solution to a set of linear ODE's (in which A is *not* a function of t) is entirely equivalent to the solution obtained if the equations are transformed to eigenspace, solved there in their uncoupled form, and then returned as a coupled set to real space.

The importance of this concept resides in its message that we can analyze time-marching methods by applying them to a single, uncoupled equation and our conclusions will apply in general. This is helpful both in analyzing the accuracy of time-marching methods and in studying their stability, topics which are covered in Chapters 6 and 7.

Our next objective is to find a "typical" single ODE to analyze. We found the uncoupled solution to a set of ODE's in Section 4.2. A typical member of the family is

$$\frac{\mathrm{d}w_m}{\mathrm{d}t} = \lambda_m w_m - g_m(t) \ . \tag{4.44}$$

The goal in our analysis is to study typical behavior of general situations, not particular problems. For such a purpose (4.44) is not quite satisfactory. The role of λ_m is clear; it stands for some representative eigenvalue in the original A matrix. However, the question is: What should we use for $g_m(t)$ when the time dependence cannot be ignored? To answer this question, we note that, in principle, one can express any one of the forcing terms $g_m(t)$ as a finite Fourier series. For example

$$-g(t) = \sum_k a_k \mathrm{e}^{\mathrm{i}kt}$$

for which (4.44) has the exact solution

$$w(t) = c\,\mathrm{e}^{\lambda t} + \sum_k \frac{a_k \mathrm{e}^{\mathrm{i}kt}}{\mathrm{i}k - \lambda} \ .$$

From this we can extract the kth term and replace $\mathrm{i}k$ with μ. This leads to:

> The representative ODE
>
> $$\frac{\mathrm{d}w}{\mathrm{d}t} = \lambda w + a\mathrm{e}^{\mu t} \ , \tag{4.45}$$

which can be used to evaluate all manner of time-marching methods. In such evaluations the parameters λ and μ must be allowed to take the worst possible combination of values that might occur in the ODE eigensystem. The exact solution of the representative ODE is (for $\mu \neq \lambda$)

$$w(t) = c\,\mathrm{e}^{\lambda t} + \frac{a\mathrm{e}^{\mu t}}{\mu - \lambda} \,. \tag{4.46}$$

Exercises

4.1 Consider the finite-difference operator derived in Exercise 3.2. Using this operator to approximate the spatial derivative in the linear convection equation, write the semi-discrete form obtained with periodic boundary conditions on a 5-point grid $(M = 5)$.

4.2 Write the semi-discrete form resulting from the application of second-order centered differences to the following equation on the domain $0 \leq x \leq 1$ with boundary conditions $u(0) = 0$, $u(1) = 1$:

$$\frac{\partial u}{\partial t} = \frac{\partial^2 u}{\partial x^2} - 6x \,.$$

4.3 Consider a grid with 10 interior points spanning the domain $0 \leq x \leq \pi$. For initial conditions $u(x,0) = \sin(mx)$ and boundary conditions $u(0,t) = u(\pi,t) = 0$, plot the exact solution of the diffusion equation with $\nu = 1$ at $t = 1$ with $m = 1$ and $m = 3$. (Plot the solution at the grid nodes only.) Calculate the corresponding modified wavenumbers for the second-order centered operator from (4.37). Calculate and plot the corresponding ODE solutions.

4.4 Consider the matrix

$$A = -B_\mathrm{p}(10: -1, 0, 1)/(2\varDelta x)$$

corresponding to the ODE form of the biconvection equation resulting from the application of second-order central differencing on a 10-point grid. Note that the domain is $0 \leq x \leq 2\pi$ and $\varDelta x = 2\pi/10$. The grid nodes are given by $x_j = j\varDelta x$, $j = 0, 1, \ldots 9$. The eigenvalues of the above matrix A, as well as the matrices X and X^{-1}, can be found from Appendix B.4. Using these, compute and plot the ODE solution at $t = 2\pi$ for the initial condition $u(x,0) = \sin x$. Compare with the exact solution of the PDE. Calculate the numerical phase speed from the modified wavenumber corresponding to this initial condition and show that it is consistent with the ODE solution. Repeat for the initial condition $u(x,0) = \sin 2x$.

5. Finite-Volume Methods

In Chapter 3, we saw how to derive finite-difference approximations to arbitrary derivatives. In Chapter 4, we saw that the application of a finite-difference approximation to the spatial derivatives in our model PDE's produces a coupled set of ODE's. In this chapter, we will show how similar semi-discrete forms can be derived using finite-volume approximations in space. Finite-volume methods have become popular in CFD as a result, primarily, of two advantages. First, they ensure that the discretization is conservative, i.e. mass, momentum, and energy are conserved in a discrete sense. While this property can usually be obtained using a finite-difference formulation, it is obtained naturally from a finite-volume formulation. Second, finite-volume methods do not require a coordinate transformation in order to be applied on irregular meshes. As a result, they can be applied on *unstructured* meshes consisting of arbitrary polyhedra in three dimensions or arbitrary polygons in two dimensions. This increased flexibility can be advantageous in generating grids about complex geometries.

Finite-volume methods are applied to the integral form of the governing equations, either in the form of (2.1) or (2.2). Consistent with our emphasis on the semi-discrete approach, we will study the latter form, which is

$$\frac{\mathrm{d}}{\mathrm{d}t} \int_{V(t)} Q \mathrm{d}V + \oint_{S(t)} \mathbf{n} \cdot \mathbf{F} \mathrm{d}S = \int_{V(t)} P \mathrm{d}V \ . \qquad (5.1)$$

We will begin by presenting the basic concepts which apply to finite-volume strategies. Next we will give our model equations in the form of (5.1). This will be followed by several examples which hopefully make these concepts clear.

5.1 Basic Concepts

The basic idea of a finite-volume method is to satisfy the integral form of the conservation law to some degree of approximation for each of many contiguous control volumes which cover the domain of interest. Thus the volume V in (5.1) is that of a control volume whose shape is dependent on the nature of the grid. In our examples, we will consider only control volumes which do not vary with time. Examining (5.1), we see that several approximations must be made. The flux is required at the boundary of the control volume, which is a closed surface in three dimensions and a closed contour in two dimensions. This flux must then be integrated to find the net flux through the boundary. Similarly, the source term P must be integrated over the control volume. Next a time-marching method[1] can be applied to find the value of

$$\int_V Q \mathrm{d}V \qquad (5.2)$$

at the next time step.

Let us consider these approximations in more detail. First, we note that the average value of Q in a cell with volume V is

$$\bar{Q} \equiv \frac{1}{V} \int_V Q \mathrm{d}V \qquad (5.3)$$

and (5.1) can be written as

$$V \frac{\mathrm{d}}{\mathrm{d}t} \bar{Q} + \oint_S \mathbf{n} \cdot \mathbf{F} \mathrm{d}S = \int_V P \mathrm{d}V \qquad (5.4)$$

for a control volume which does not vary with time. Thus after applying a time-marching method, we have updated values of the cell-averaged quantities $\bar{Q}$. In order to evaluate the fluxes, which are a function of Q, at the control-volume boundary, Q can be represented within the cell by some piecewise approximation which produces the correct value of $\bar{Q}$. This is a form of interpolation often referred to as *reconstruction*. As we shall see in our examples, each cell will have a different piecewise approximation to Q. When these are used to calculate $\mathbf{F}(Q)$, they will generally produce different approximations to the flux at the boundary between two control volumes, that is, the flux will be discontinuous. A nondissipative scheme analogous to centered differencing is obtained by taking the average of these two fluxes. Another approach known as flux-difference splitting is described in Chapter 11.

The basic elements of a finite-volume method are thus the following:

(1) Given the value of $\bar{Q}$ for each control volume, construct an approximation to $Q(x, y, z)$ in each control volume. Using this approximation, find Q at the control-volume boundary. Evaluate $\mathbf{F}(Q)$ at the boundary. Since there is a distinct approximation to $Q(x, y, z)$ in each control volume, two distinct values of the flux will generally be obtained at any point on the boundary between two control volumes.

(2) Apply some strategy for resolving the discontinuity in the flux at the control-volume boundary to produce a single value of $\mathbf{F}(Q)$ at any point on the boundary. This issue is discussed in Section 11.4.2.

---

[1] Time-marching methods will be discussed in the next chapter.

(3) Integrate the flux to find the net flux through the control-volume boundary using some sort of quadrature.

(4) Advance the solution in time to obtain new values of $\bar{Q}$.

The order of accuracy of the method is dependent on each of the approximations. These ideas should be clarified by the examples in the remainder of this chapter.

In order to include diffusive fluxes, the following relation between ∇Q and Q is sometimes used:

$$\int_V \nabla Q dV = \oint_S \mathbf{n} Q dS \tag{5.5}$$

or, in two dimensions,

$$\int_A \nabla Q dA = \oint_C \mathbf{n} Q dl , \tag{5.6}$$

where the unit vector $\mathbf{n}$ points outward from the surface or contour.

5.2 Model Equations in Integral Form

5.2.1 The Linear Convection Equation

A two-dimensional form of the linear convection equation can be written as

$$\frac{\partial u}{\partial t} + a \cos\theta \frac{\partial u}{\partial x} + a \sin\theta \frac{\partial u}{\partial y} = 0 . \tag{5.7}$$

This PDE governs a simple plane wave convecting the scalar quantity, $u(x, y, t)$ with speed a along a straight line making an angle θ with respect to the x-axis. The one-dimensional form is recovered with $\theta = 0$.

For unit speed a, the two-dimensional linear convection equation is obtained from the general divergence form, (2.3), with

$$Q = u \tag{5.8}$$
$$\mathbf{F} = \mathbf{i} u \cos\theta + \mathbf{j} u \sin\theta \tag{5.9}$$
$$P = 0 . \tag{5.10}$$

Since Q is a scalar, $\mathbf{F}$ is simply a vector. Substituting these expressions into a two-dimensional form of (2.2) gives the following integral form

$$\frac{d}{dt} \int_A u dA + \oint_C \mathbf{n} . (\mathbf{i} u \cos\theta + \mathbf{j} u \sin\theta) dl = 0 , \tag{5.11}$$

where A is the area of the cell which is bounded by the closed contour C.

5.2.2 The Diffusion Equation

The integral form of the two-dimensional diffusion equation with no source term and unit diffusion coefficient ν is obtained from the general divergence form, (2.3), with

$$Q = u \tag{5.12}$$

$$\mathbf{F} = -\nabla u \tag{5.13}$$

$$= -\left(\mathbf{i}\frac{\partial u}{\partial x} + \mathbf{j}\frac{\partial u}{\partial y} \right) \tag{5.14}$$

$$P = 0 . \tag{5.15}$$

Using these, we find

$$\frac{d}{dt}\int_A u \, dA = \oint_C \mathbf{n} \cdot \left(\mathbf{i}\frac{\partial u}{\partial x} + \mathbf{j}\frac{\partial u}{\partial y} \right) dl \tag{5.16}$$

to be the integral form of the two-dimensional diffusion equation.

5.3 One-Dimensional Examples

We restrict our attention to a scalar dependent variable u and a scalar flux f, as in the model equations. We consider an equispaced grid with spacing Δx. The nodes of the grid are located at $x_j = j\Delta x$ as usual. Control volume j extends from $x_j - \Delta x/2$ to $x_j + \Delta x/2$, as shown in Figure 5.1. We will use the following notation:

$$x_{j-1/2} = x_j - \Delta x/2, \qquad x_{j+1/2} = x_j + \Delta x/2 , \tag{5.17}$$

$$u_{j\pm1/2} = u(x_{j\pm1/2}), \qquad f_{j\pm1/2} = f(u_{j\pm1/2}) . \tag{5.18}$$

With these definitions, the cell-average value becomes

$$\bar{u}_j(t) \equiv \frac{1}{\Delta x}\int_{x_{j-1/2}}^{x_{j+1/2}} u(x,t)\,dx \tag{5.19}$$

and the integral form becomes

$$\frac{d}{dt}(\Delta x \bar{u}_j) + f_{j+1/2} - f_{j-1/2} = \int_{x_{j-1/2}}^{x_{j+1/2}} P\,dx . \tag{5.20}$$

Now with $\xi = x - x_j$, we can expand $u(x)$ in (5.19) in a Taylor series about x_j (with t fixed) to get

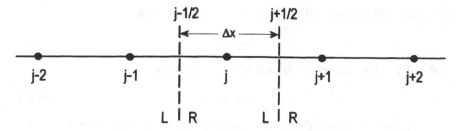

Fig. 5.1. Control volume in one dimension

$$\bar{u}_j \equiv \frac{1}{\Delta x} \int_{-\Delta x/2}^{\Delta x/2} \left[u_j + \xi \left(\frac{\partial u}{\partial x} \right)_j + \frac{\xi^2}{2} \left(\frac{\partial^2 u}{\partial x^2} \right)_j + \frac{\xi^3}{6} \left(\frac{\partial^3 u}{\partial x^3} \right)_j + \dots \right] d\xi$$

$$= u_j + \frac{\Delta x^2}{24} \left(\frac{\partial^2 u}{\partial x^2} \right)_j + \frac{\Delta x^4}{1920} \left(\frac{\partial^4 u}{\partial x^4} \right)_j + O(\Delta x^6) \tag{5.21}$$

or

$$\bar{u}_j = u_j + O(\Delta x^2) , \tag{5.22}$$

where u_j is the value at the center of the cell. Hence the cell-average value and the value at the center of the cell differ by a term of second order.

5.3.1 A Second-Order Approximation to the Convection Equation

In one dimension, the integral form of the linear convection equation, (5.11), becomes

$$\Delta x \frac{d\bar{u}_j}{dt} + f_{j+1/2} - f_{j-1/2} = 0 \tag{5.23}$$

with $f = u$. We choose a piecewise-constant approximation to $u(x)$ in each cell such that

$$u(x) = \bar{u}_j \qquad x_{j-1/2} \le x \le x_{j+1/2} . \tag{5.24}$$

Evaluating this at $j + 1/2$ gives

$$f_{j+1/2}^{L} = f(u_{j+1/2}^{L}) = u_{j+1/2}^{L} = \bar{u}_j , \tag{5.25}$$

where the L indicates that this approximation to $f_{j+1/2}$ is obtained from the approximation to $u(x)$ in the cell to the *left* of $x_{j+1/2}$, as shown in Figure 5.1. The cell to the *right* of $x_{j+1/2}$, which is cell $j + 1$, gives

$$f_{j+1/2}^{R} = \bar{u}_{j+1} . \tag{5.26}$$

Similarly, cell j is the cell to the right of $x_{j-1/2}$, giving

$$f^{\text{R}}_{j-1/2} = \bar{u}_j \tag{5.27}$$

and cell $j-1$ is the cell to the left of $x_{j-1/2}$, giving

$$f^{\text{L}}_{j-1/2} = \bar{u}_{j-1} . \tag{5.28}$$

We have now accomplished the first step from the list in Section 5.1; we have defined the fluxes at the cell boundaries in terms of the cell-average data. In this example, the discontinuity in the flux at the cell boundary is resolved by taking the average of the fluxes on either side of the boundary. Thus

$$\hat{f}_{j+1/2} = \frac{1}{2}(f^{\text{L}}_{j+1/2} + f^{\text{R}}_{j+1/2}) = \frac{1}{2}(\bar{u}_j + \bar{u}_{j+1}) \tag{5.29}$$

and

$$\hat{f}_{j-1/2} = \frac{1}{2}(f^{\text{L}}_{j-1/2} + f^{\text{R}}_{j-1/2}) = \frac{1}{2}(\bar{u}_{j-1} + \bar{u}_j) , \tag{5.30}$$

where $\hat{f}$ denotes a *numerical* flux which is an approximation to the exact flux.

Substituting (5.29) and (5.30) into the integral form, (5.23), we obtain

$$\Delta x \frac{\mathrm{d}\bar{u}_j}{\mathrm{d}t} + \frac{1}{2}(\bar{u}_j + \bar{u}_{j+1}) - \frac{1}{2}(\bar{u}_{j-1} + \bar{u}_j) = \Delta x \frac{\mathrm{d}\bar{u}_j}{\mathrm{d}t} + \frac{1}{2}(\bar{u}_{j+1} - \bar{u}_{j-1}) = 0 . \tag{5.31}$$

With periodic boundary conditions, this point operator produces the following semi-discrete form:

$$\frac{\mathrm{d}\vec{\bar{u}}}{\mathrm{d}t} = -\frac{1}{2\Delta x} B_{\text{p}}(-1,0,1)\vec{\bar{u}} \tag{5.32}$$

This is identical to the expression obtained using second-order centered differences, except it is written in terms of the cell average $\bar{u}$, rather than the nodal values, $\vec{u}$. Hence our analysis and understanding of the eigensystem of the matrix $B_{\text{p}}(-1,0,1)$ is relevant to finite-volume methods as well as finite-difference methods. Since the eigenvalues of $B_{\text{p}}(-1,0,1)$ are pure imaginary, we can conclude that the use of the average of the fluxes on either side of the cell boundary, as in (5.29) and (5.30), can lead to a nondissipative finite-volume method.

5.3.2 A Fourth-Order Approximation to the Convection Equation

Let us replace the piecewise-constant approximation in Section 5.3.1 with a piecewise-quadratic approximation as follows

$$u(\xi) = a\xi^2 + b\xi + c \,, \tag{5.33}$$

where ξ is again equal to $x - x_j$. The three parameters a, b, and c are chosen to satisfy the following constraints:

$$\frac{1}{\Delta x} \int_{-3\Delta x/2}^{-\Delta x/2} u(\xi)d\xi = \bar{u}_{j-1}$$

$$\frac{1}{\Delta x} \int_{-\Delta x/2}^{\Delta x/2} u(\xi)d\xi = \bar{u}_j \tag{5.34}$$

$$\frac{1}{\Delta x} \int_{\Delta x/2}^{3\Delta x/2} u(\xi)d\xi = \bar{u}_{j+1} \,.$$

These constraints lead to

$$a = \frac{\bar{u}_{j+1} - 2\bar{u}_j + \bar{u}_{j-1}}{2\Delta x^2}$$

$$b = \frac{\bar{u}_{j+1} - \bar{u}_{j-1}}{2\Delta x} \tag{5.35}$$

$$c = \frac{-\bar{u}_{j-1} + 26\bar{u}_j - \bar{u}_{j+1}}{24} \,.$$

With these values of a, b, and c, the piecewise-quadratic approximation produces the following values at the cell boundaries:

$$u_{j+1/2}^{\mathrm{L}} = \frac{1}{6}(2\bar{u}_{j+1} + 5\bar{u}_j - \bar{u}_{j-1}) \tag{5.36}$$

$$u_{j-1/2}^{\mathrm{R}} = \frac{1}{6}(-\bar{u}_{j+1} + 5\bar{u}_j + 2\bar{u}_{j-1}) \tag{5.37}$$

$$u_{j+1/2}^{\mathrm{R}} = \frac{1}{6}(-\bar{u}_{j+2} + 5\bar{u}_{j+1} + 2\bar{u}_j) \tag{5.38}$$

$$u_{j-1/2}^{\mathrm{L}} = \frac{1}{6}(2\bar{u}_j + 5\bar{u}_{j-1} - \bar{u}_{j-2}) \tag{5.39}$$

using the notation defined in Section 5.3.1. Recalling that $f = u$, we again use the average of the fluxes on either side of the boundary to obtain

$$\hat{f}_{j+1/2} = \frac{1}{2}[f(u_{j+1/2}^{\mathrm{L}}) + f(u_{j+1/2}^{\mathrm{R}})]$$

$$= \frac{1}{12}(-\bar{u}_{j+2} + 7\bar{u}_{j+1} + 7\bar{u}_j - \bar{u}_{j-1}) \tag{5.40}$$

and

$$\hat{f}_{j-1/2} = \frac{1}{2}[f(u^L_{j-1/2}) + f(u^R_{j-1/2})]$$

$$= \frac{1}{12}(-\bar{u}_{j+1} + 7\bar{u}_j + 7\bar{u}_{j-1} - \bar{u}_{j-2}) \ . \tag{5.41}$$

Substituting these expressions into the integral form, (5.23), gives

$$\Delta x \frac{d\bar{u}_j}{dt} + \frac{1}{12}(-\bar{u}_{j+2} + 8\bar{u}_{j+1} - 8\bar{u}_{j-1} + \bar{u}_{j-2}) = 0 \ . \tag{5.42}$$

This is a fourth-order approximation to the *integral* form of the equation, as can be verified using Taylor series expansions. With periodic boundary conditions, the following semi-discrete form is obtained:

$$\frac{d\vec{\bar{u}}}{dt} = -\frac{1}{12\Delta x} B_p(1, -8, 0, 8, -1)\vec{\bar{u}} \ . \tag{5.43}$$

This is a system of ODE's governing the evolution of the cell-average data.

5.3.3 A Second-Order Approximation to the Diffusion Equation

In this section, we describe two approaches to deriving a finite-volume approximation to the diffusion equation. The first approach is simpler to extend to multidimensions, while the second approach is more suited to extension to higher-order accuracy.

In one dimension, the integral form of the diffusion equation, (5.16), becomes

$$\Delta x \frac{d\bar{u}_j}{dt} + f_{j+1/2} - f_{j-1/2} = 0 \tag{5.44}$$

with $f = -\nabla u = -\partial u / \partial x$. Also, (5.6) becomes

$$\int_a^b \frac{\partial u}{\partial x} dx = u(b) - u(a) \ . \tag{5.45}$$

We can thus write the following expression for the average value of the gradient of u over the interval $x_j \leq x \leq x_{j+1}$:

$$\frac{1}{\Delta x} \int_{x_j}^{x_{j+1}} \frac{\partial u}{\partial x} dx = \frac{1}{\Delta x}(u_{j+1} - u_j) \ . \tag{5.46}$$

From (5.22), we know that the value of a continuous function at the center of a given interval is equal to the average value of the function over the interval to second-order accuracy. Hence, to second-order, we can write

$$\hat{f}_{j+1/2} = -\left(\frac{\partial u}{\partial x}\right)_{j+1/2} = -\frac{1}{\Delta x}(\bar{u}_{j+1} - \bar{u}_j) \ . \tag{5.47}$$

Similarly,

$$\hat{f}_{j-1/2} = -\frac{1}{\Delta x}(\bar{u}_j - \bar{u}_{j-1}) . \tag{5.48}$$

Substituting these into the integral form, (5.44), we obtain

$$\Delta x \frac{d\bar{u}_j}{dt} = \frac{1}{\Delta x}(\bar{u}_{j-1} - 2\bar{u}_j + \bar{u}_{j+1}) \tag{5.49}$$

or, with Dirichlet boundary conditions,

$$\frac{d\vec{\bar{u}}}{dt} = \frac{1}{\Delta x^2}B(1, -2, 1)\vec{\bar{u}} + \left(\vec{bc}\right) . \tag{5.50}$$

This provides a semi-discrete finite-volume approximation to the diffusion equation, and we see that the properties of the matrix $B(1, -2, 1)$ are relevant to the study of finite-volume methods as well as finite-difference methods.

For our second approach, we use a piecewise-quadratic approximation as in Section 5.3.2. From (5.33) we have

$$\frac{\partial u}{\partial x} = \frac{\partial u}{\partial \xi} = 2a\xi + b \tag{5.51}$$

with a and b given in (5.35). With $f = -\partial u/\partial x$, this gives

$$f^{\mathrm{R}}_{j+1/2} = f^{\mathrm{L}}_{j+1/2} = -\frac{1}{\Delta x}(\bar{u}_{j+1} - \bar{u}_j) \tag{5.52}$$

$$f^{\mathrm{R}}_{j-1/2} = f^{\mathrm{L}}_{j-1/2} = -\frac{1}{\Delta x}(\bar{u}_j - \bar{u}_{j-1}) . \tag{5.53}$$

Notice that there is no discontinuity in the flux at the cell boundary. This produces

$$\frac{d\bar{u}_j}{dt} = \frac{1}{\Delta x^2}(\bar{u}_{j-1} - 2\bar{u}_j + \bar{u}_{j+1}) , \tag{5.54}$$

which is identical to (5.49). The resulting semi-discrete form with periodic boundary conditions is

$$\frac{d\vec{\bar{u}}}{dt} = \frac{1}{\Delta x^2}B_{\mathrm{p}}(1, -2, 1)\vec{\bar{u}} , \tag{5.55}$$

which is written entirely in terms of cell-average data.

5.4 A Two-Dimensional Example

The above one-dimensional examples of finite-volume approximations obscure some of the practical aspects of such methods. Thus our final example is a finite-volume approximation to the two-dimensional linear convection equation on a grid consisting of regular triangles, as shown in Figure 5.2. As in Section 5.3.1, we use a piecewise-constant approximation in each control volume, and the flux at the control volume boundary is the average of the fluxes obtained on either side of the boundary. The nodal data are stored at the vertices of the triangles formed by the grid. The control volumes are regular hexagons with area A, Δ is the length of the sides of the triangles, and ℓ is the length of the sides of the hexagons. The following relations hold between ℓ, Δ, and A:

$$\ell = \frac{1}{\sqrt{3}}\Delta$$
$$A = \frac{3\sqrt{3}}{2}\ell^2$$
$$\frac{\ell}{A} = \frac{2}{3\Delta} \ . \tag{5.56}$$

The two-dimensional form of a conservation law is

$$\frac{d}{dt}\int_A Q dA + \oint_C \mathbf{n}\cdot\mathbf{F}dl = 0 \ , \tag{5.57}$$

where we have ignored the source term. The contour in the line integral is composed of the sides of the hexagon. Since these sides are all straight, the

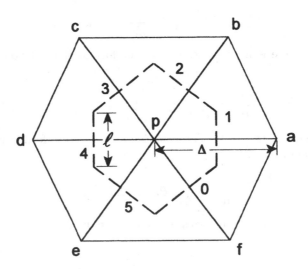

Fig. 5.2. Triangular grid

unit normals can be taken outside the integral and the flux balance is given by

$$\frac{d}{dt}\int_A Q\, dA + \sum_{\nu=0}^{5} \mathbf{n}_\nu \cdot \int_\nu \mathbf{F} dl = 0$$

where ν indexes a side of the hexagon, as shown in Figure 5.2. A list of the normals for the mesh orientation shown is given in Table 5.1.

Table 5.1. Outward normals, see Figure 5.2.
($\mathbf{i}$ and $\mathbf{j}$ are unit normals along x and y, respectively)

Side, ν	Outward Normal, $\mathbf{n}$
0	$(\mathbf{i} - \sqrt{3}\,\mathbf{j})/2$
1	$\mathbf{i}$
2	$(\mathbf{i} + \sqrt{3}\,\mathbf{j})/2$
3	$(-\mathbf{i} + \sqrt{3}\,\mathbf{j})/2$
4	$-\mathbf{i}$
5	$(-\mathbf{i} - \sqrt{3}\,\mathbf{j})/2$

For (5.11), the two-dimensional linear convection equation, we have for side ν

$$\mathbf{n}_\nu \cdot \int_\nu \mathbf{F} dl = \mathbf{n}_\nu \cdot (\mathbf{i}\cos\theta + \mathbf{j}\sin\theta) \int_{-\ell/2}^{\ell/2} u_\nu(\xi)d\xi , \qquad (5.58)$$

where ξ is a length measured from the middle of a side ν. Making the change of variable $z = \xi/\ell$, one has the expression

$$\int_{-\ell/2}^{\ell/2} u(\xi)d\xi = \ell \int_{-1/2}^{1/2} u(z)dz . \qquad (5.59)$$

Then, in terms of u and the hexagon area A, we have

$$\frac{d}{dt}\int_A u\, dA + \sum_{\nu=0}^{5} \mathbf{n}_\nu \cdot (\mathbf{i}\cos\theta + \mathbf{j}\sin\theta)\left[\ell \int_{-1/2}^{1/2} u(z)dz\right]_\nu = 0 . \qquad (5.60)$$

The values of $\mathbf{n}_\nu \cdot (\mathbf{i}\cos\theta + \mathbf{j}\sin\theta)$ are given by the expressions in Table 5.2. There are no numerical approximations in (5.60). That is, if the integrals in the equation are evaluated exactly, the integrated time rate of change of the integral of u over the area of the hexagon is known exactly.

Introducing the cell average,

$$\int_A u\, dA = A\bar{u}_p \qquad (5.61)$$

Table 5.2. Weights of flux integrals, see (5.60)

Side, ν	$\mathbf{n}_\nu \cdot (\mathbf{i}\cos\theta + \mathbf{j}\sin\theta)$
0	$(\cos\theta - \sqrt{3}\sin\theta)/2$
1	$\cos\theta$
2	$(\cos\theta + \sqrt{3}\sin\theta)/2$
3	$(-\cos\theta + \sqrt{3}\sin\theta)/2$
4	$-\cos\theta$
5	$(-\cos\theta - \sqrt{3}\sin\theta)/2$

and the piecewise-constant approximation $u = \bar{u}_p$ over the entire hexagon, the approximation to the flux integral becomes trivial. Taking the average of the flux on either side of each edge of the hexagon gives for edge 1:

$$\int_1 u(z)\mathrm{d}z = \frac{\bar{u}_p + \bar{u}_a}{2} \int_{-1/2}^{1/2} \mathrm{d}z = \frac{\bar{u}_p + \bar{u}_a}{2} . \tag{5.62}$$

Similarly, we have for the other five edges

$$\int_2 u(z)\mathrm{d}z = \frac{\bar{u}_p + \bar{u}_b}{2} \tag{5.63}$$

$$\int_3 u(z)\mathrm{d}z = \frac{\bar{u}_p + \bar{u}_c}{2} \tag{5.64}$$

$$\int_4 u(z)\mathrm{d}z = \frac{\bar{u}_p + \bar{u}_d}{2} \tag{5.65}$$

$$\int_5 u(z)\mathrm{d}z = \frac{\bar{u}_p + \bar{u}_e}{2} \tag{5.66}$$

$$\int_0 u(z)\mathrm{d}z = \frac{\bar{u}_p + \bar{u}_f}{2} . \tag{5.67}$$

Substituting these into (5.60), along with the expressions in Table 5.2, we obtain

$$A\frac{\mathrm{d}\bar{u}_p}{\mathrm{d}t} + \frac{\ell}{4}[(2\cos\theta)(\bar{u}_a - \bar{u}_d) + (\cos\theta + \sqrt{3}\sin\theta)(\bar{u}_b - \bar{u}_e)$$
$$+ (-\cos\theta + \sqrt{3}\sin\theta)(\bar{u}_c - \bar{u}_f)] = 0 \tag{5.68}$$

or

$$\frac{\mathrm{d}\bar{u}_p}{\mathrm{d}t} + \frac{1}{6\Delta}[(2\cos\theta)(\bar{u}_a - \bar{u}_d) + (\cos\theta + \sqrt{3}\sin\theta)(\bar{u}_b - \bar{u}_e)$$
$$+ (-\cos\theta + \sqrt{3}\sin\theta)(\bar{u}_c - \bar{u}_f)] = 0 . \tag{5.69}$$

The reader can verify, using Taylor series expansions, that this is a second-order approximation to the integral form of the two-dimensional linear convection equation.

Exercises

5.1 Use Taylor series to verify that (5.42) is a fourth-order approximation to (5.23).

5.2 Find the semi-discrete ODE form governing the cell-average data resulting from the use of a linear approximation in developing a finite-volume method for the linear convection equation. Use the following linear approximation:

$$u(\xi) = a\xi + b ,$$

where $b = \bar{u}_j$ and

$$a = \frac{\bar{u}_{j+1} - \bar{u}_{j-1}}{2\Delta x}$$

and use the average flux at the cell interface.

5.3 Using the first approach given in Section 5.3.3, derive a finite-volume approximation to the spatial terms in the two-dimensional diffusion equation on a square grid.

5.4 Repeat Exercise 5.3 for a grid consisting of equilateral triangles.

6. Time-Marching Methods for ODE'S

After discretizing the spatial derivatives in the governing PDE's (such as the Navier–Stokes equations), we obtain a coupled system of nonlinear ODE's in the form

$$\frac{d\vec{u}}{dt} = \vec{F}(\vec{u}, t) .\tag{6.1}$$

These can be integrated in time using a time-marching method to obtain a time-accurate solution to an *unsteady* flow problem. For a *steady* flow problem, spatial discretization leads to a coupled system of nonlinear algebraic equations in the form

$$\vec{F}(\vec{u}) = 0 .\tag{6.2}$$

As a result of the nonlinearity of these equations, some sort of iterative method is required to obtain a solution. For example, one can consider the use of Newton's method, which is widely used for nonlinear algebraic equations (see Section 6.10.3). This produces an iterative method in which a coupled system of linear algebraic equations must be solved at each iteration. These can be solved iteratively using relaxation methods, which will be discussed in Chapter 9, or directly using Gaussian elimination or some variation thereof.

Alternatively, one can consider a time-dependent path to the steady state and use a time-marching method to integrate the unsteady form of the equations until the solution is sufficiently close to the steady solution. The subject of the present chapter, time-marching methods for ODE's, is thus relevant to both steady and unsteady flow problems. When using a time-marching method to compute steady flows, the goal is simply to remove the transient portion of the solution as quickly as possible; time-accuracy is not required. This motivates the study of stability and stiffness, topics which are covered in the next two chapters.

Application of a spatial discretization to a PDE produces a coupled system of ODE's. Application of a time-marching method to an ODE produces an ordinary *difference* equation (OΔE). In earlier chapters, we developed exact solutions to our model PDE's and ODE's. In this chapter we will present some basic theory of linear OΔE's which closely parallels that for linear ODE's, and, using this theory, we will develop exact solutions for the model OΔE's arising from the application of time-marching methods to the model ODE's.

6.1 Notation

Using the semi-discrete approach, we reduce our PDE to a set of coupled ODE's represented in general by (4.1). However, for the purpose of this chapter, we need only consider the scalar case

$$\frac{du}{dt} = u' = F(u, t) \ . \tag{6.3}$$

Although we use u to represent the dependent variable, rather than w, the reader should recall the arguments made in Chapter 4 to justify the study of a scalar ODE. Our first task is to find numerical approximations that can be used to carry out the time integration of (6.3) to some given accuracy, where accuracy can be measured either in a local or a global sense. We then face a further task concerning the numerical stability of the resulting methods, but we postpone such considerations to the next chapter.

In Chapter 2, we introduced the convention that the n subscript, or the (n) superscript, always points to a discrete time value, and h represents the time interval Δt. Combining this notation with (6.3) gives

$$u'_n = F_n = F(u_n, t_n) \quad , \quad t_n = nh \ .$$

Often we need a more sophisticated notation for intermediate time steps involving temporary calculations denoted by $\tilde{u}$, $\bar{u}$, etc. For these we use the notation

$$\tilde{u}'_{n+\alpha} = \tilde{F}_{n+\alpha} = F(\tilde{u}_{n+\alpha}, t_n + \alpha h) \ .$$

The choice of u' or F to express the derivative in a scheme is arbitrary. They are both commonly used in the literature on ODE's.

The methods we study are to be applied to linear or nonlinear ODE's, but the methods themselves are formed by linear combinations of the dependent variable and its derivative at various time intervals. They are represented conceptually by

$$u_{n+1} = f\left(\beta_1 u'_{n+1}, \beta_0 u'_n, \beta_{-1} u'_{n-1}, \cdots, \alpha_0 u_n, \alpha_{-1} u_{n-1}, \cdots\right) \ . \tag{6.4}$$

With an appropriate choice of the α's and β's, these methods can be constructed to give a local Taylor series accuracy of any order. The methods are said to be *explicit* if $\beta_1 = 0$ and *implicit* otherwise. An *explicit* method is one in which the new predicted solution is only a function of known data, for example, u'_n, u'_{n-1}, u_n, and u_{n-1} for a method using two previous time levels, and therefore the time advance is simple. For an *implicit* method, the new predicted solution is also a function of the time derivative at the new time level, that is, u'_{n+1}. As we shall see, for systems of ODE's and nonlinear problems, *implicit* methods require more complicated strategies to solve for u_{n+1} than *explicit* methods.

6.2 Converting Time-Marching Methods to OΔE's

Examples of some very common forms of methods used for time-marching general ODE's are:

$$u_{n+1} = u_n + hu'_n \tag{6.5}$$

$$u_{n+1} = u_n + hu'_{n+1} \tag{6.6}$$

and

$$\tilde{u}_{n+1} = u_n + hu'_n$$
$$u_{n+1} = \frac{1}{2}(u_n + \tilde{u}_{n+1} + h\tilde{u}'_{n+1}) \,. \tag{6.7}$$

According to the conditions presented under (6.4), the first and third of these are examples of explicit methods. We refer to them as the explicit Euler method and the MacCormack predictor–corrector method,[1] respectively. The second is implicit and referred to as the implicit (or backward) Euler method.

These methods are simple recipes for the time advance of a function in terms of its value and the value of its derivative, at given time intervals. The material presented in Chapter 4 develops a basis for evaluating such methods by introducing the concept of the representative equation

$$\frac{du}{dt} = u' = \lambda u + ae^{\mu t} \tag{6.8}$$

written here in terms of the dependent variable, u. The value of this equation arises from the fact that, by applying a time-marching method, we can analytically convert such a linear ODE into a linear OΔE . The latter are subject to a whole body of analysis that is similar in many respects to, and just as powerful as, the theory of ODE's. We next consider examples of this conversion process and then go into the general theory on solving OΔE's.

Apply the simple explicit Euler scheme, (6.5), to (6.8). There results

$$u_{n+1} = u_n + h(\lambda u_n + ae^{\mu hn})$$

or

$$u_{n+1} - (1 + \lambda h)u_n = hae^{\mu hn} \,. \tag{6.9}$$

Equation (6.9) is a linear OΔE , with constant coefficients, expressed in terms of the dependent variable u_n and the independent variable n. As another example, applying the implicit Euler method, (6.6), to (6.8), we find

---

[1] Here we give only MacCormack's *time-marching* method. The method commonly referred to as MacCormack's method, which is a fully-discrete method, will be presented in Section 11.3.

$$u_{n+1} = u_n + h\Big(\lambda u_{n+1} + ae^{\mu h(n+1)}\Big)$$

or

$$(1 - \lambda h)u_{n+1} - u_n = he^{\mu h} \cdot ae^{\mu hn} . \qquad (6.10)$$

As a final example, the predictor–corrector sequence, (6.7), gives

$$\tilde{u}_{n+1} - (1 + \lambda h)u_n = ahe^{\mu hn}$$

$$-\frac{1}{2}(1 + \lambda h)\tilde{u}_{n+1} + u_{n+1} - \frac{1}{2}u_n = \frac{1}{2}ahe^{\mu h(n+1)} , \qquad (6.11)$$

which is a coupled set of linear $O\Delta E$'s with constant coefficients. Note that the first line in (6.11) is identical to (6.9), since the predictor step in (6.7) is simply the explicit Euler method. The second line in (6.11) is obtained by noting that

$$\tilde{u}'_{n+1} = F(\tilde{u}_{n+1}, t_n + h)$$

$$= \lambda \tilde{u}_{n+1} + ae^{\mu h(n+1)} . \qquad (6.12)$$

Now we need to develop techniques for analyzing these difference equations so that we can compare the merits of the time-marching methods that generated them.

6.3 Solution of Linear $O\Delta E$'s with Constant Coefficients

The techniques for solving *linear difference equations* with constant coefficients is as well developed as that for ODE's, and the theory follows a remarkably parallel path. This is demonstrated by repeating some of the developments in Section 4.2, but for difference rather than differential equations.

6.3.1 First- and Second-Order Difference Equations

First-Order Equations. The simplest nonhomogeneous $O\Delta E$ of interest is given by the single first-order equation

$$u_{n+1} = \sigma u_n + ab^n , \qquad (6.13)$$

where σ, a, and b are, in general, complex parameters. The independent variable is now n rather than t, and since the equations are linear and have constant coefficients, σ is not a function of either n or u. The exact solution of (6.13) is

$$u_n = c_1 \sigma^n + \frac{ab^n}{b - \sigma} ,$$

where c_1 is a constant determined by the initial conditions. In terms of the initial value of u it can be written

$$u_n = u_0 \sigma^n + a \frac{b^n - \sigma^n}{b - \sigma} .$$

Just as in the development of (4.10), one can readily show that the solution of the defective case, $(b = \sigma)$,

$$u_{n+1} = \sigma u_n + a \sigma^n$$

is

$$u_n = \left[u_0 + a n \sigma^{-1} \right] \sigma^n .$$

This can all be easily verified by substitution.

Second-Order Equations. The homogeneous form of a second-order difference equation is given by

$$u_{n+2} + a_1 u_{n+1} + a_0 u_n = 0 . \tag{6.14}$$

Instead of the differential operator $D \equiv d/dt$ used for ODE's, we use for OΔE's the difference operator E (commonly referred to as the displacement or shift operator) and defined formally by the relations

$$u_{n+1} = E u_n \quad , \quad u_{n+k} = E^k u_n .$$

Further notice that the displacement operator also applies to exponents, thus

$$b^\alpha \cdot b^n = b^{n+\alpha} = E^\alpha \cdot b^n ,$$

where α can be any fraction or irrational number.

The roles of D and E are the same insofar as once they have been introduced to the basic equations, the value of $u(t)$ or u_n can be factored out. Thus (6.14) can now be re-expressed in an operational notation as

$$(E^2 + a_1 E + a_0) u_n = 0 , \tag{6.15}$$

which must be zero for all u_n. Equation (6.15) is known as the *operational form* of (6.14). The operational form contains a characteristic polynomial $P(E)$ which plays the same role for difference equations that $P(D)$ played for differential equations; that is, its roots determine the solution to the OΔE. In the analysis of OΔE's, we label these roots σ_1, σ_2, $\cdots$, etc, and refer to them as the *σ-roots*. They are found by solving the equation $P(\sigma) = 0$. In the simple example given above, there are just two σ roots, and in terms of them the solution can be written

$$u_n = c_1 (\sigma_1)^n + c_2 (\sigma_2)^n , \tag{6.16}$$

where c_1 and c_2 depend upon the initial conditions. The fact that (6.16) is a solution to (6.14) for all c_1, c_2, and n should be verified by substitution.

6.3.2 Special Cases of Coupled First-Order Equations

A Complete System. Coupled, first-order, linear, homogeneous difference equations have the form

$$u_1^{(n+1)} = c_{11}u_1^{(n)} + c_{12}u_2^{(n)} \qquad (6.17)$$
$$u_2^{(n+1)} = c_{21}u_1^{(n)} + c_{22}u_2^{(n)} \ ,$$

which can also be written as

$$\vec{u}_{n+1} = C\vec{u}_n \ , \quad \vec{u}_n = \left[u_1^{(n)}, u_2^{(n)}\right]^T \ , \quad C = \begin{bmatrix} c_{11} & c_{12} \\ c_{21} & c_{22} \end{bmatrix} \ .$$

The operational form of (6.17) can be written as

$$\begin{bmatrix} (c_{11} - E) & c_{12} \\ c_{21} & (c_{22} - E) \end{bmatrix} \begin{bmatrix} u_1 \\ u_2 \end{bmatrix}^{(n)} = [C - E\,I]\vec{u}_n = 0 \ ,$$

which must be zero for all u_1 and u_2. Again we are led to a characteristic polynomial, this time having the form $P(E) = \det[C - E\,I]$. The σ-roots are found from

$$P(\sigma) = \det \begin{bmatrix} (c_{11} - \sigma) & c_{12} \\ c_{21} & (c_{22} - \sigma) \end{bmatrix} = 0 \ .$$

Obviously, the σ_k are the eigenvalues of C, and following the logic of Section 4.2, if $\vec{x}$ are its eigenvectors, the solution of (6.17) is

$$\vec{u}_n = \sum_{k=1}^{2} c_k(\sigma_k)^n \vec{x}_k \ ,$$

where c_k are constants determined by the initial conditions.

A Defective System. The solution of $O\varDelta E$'s with defective eigensystems follows closely the logic in Section 4.2.2 for defective ODE's. For example, one can show that the solution to

$$\begin{bmatrix} \bar{u}_{n+1} \\ \hat{u}_{n+1} \\ u_{n+1} \end{bmatrix} = \begin{bmatrix} \sigma & & \\ 1 & \sigma & \\ & 1 & \sigma \end{bmatrix} \begin{bmatrix} \bar{u}_n \\ \hat{u}_n \\ u_n \end{bmatrix}$$

is

$$\bar{u}_n = \bar{u}_0\sigma^n$$
$$\hat{u}_n = \left[\hat{u}_0 + \bar{u}_0 n\sigma^{-1}\right]\sigma^n \qquad (6.18)$$
$$u_n = \left[u_0 + \hat{u}_0 n\sigma^{-1} + \bar{u}_0 \frac{n(n-1)}{2}\sigma^{-2}\right]\sigma^n \ .$$

6.4 Solution of the Representative OΔE's

6.4.1 The Operational Form and its Solution

Examples of the nonhomogeneous, linear, first-order ordinary difference equations produced by applying a time-marching method to the representative equation are given by (6.9)–(6.11). Using the displacement operator, E, these equations can be written

$$[E - (1 + \lambda h)]u_n = h \cdot ae^{\mu h n} \tag{6.19}$$

$$[(1 - \lambda h)E - 1]u_n = h \cdot E \cdot ae^{\mu h n} \tag{6.20}$$

$$\begin{bmatrix} E & -(1 + \lambda h) \\ -\frac{1}{2}(1 + \lambda h)E & E - \frac{1}{2} \end{bmatrix} \begin{bmatrix} \tilde{u} \\ u \end{bmatrix}_n = h \cdot \begin{bmatrix} 1 \\ \frac{1}{2}E \end{bmatrix} \cdot ae^{\mu h n} . \tag{6.21}$$

All three of these equations are subsets of the *operational form* of the representative OΔE

$$\boxed{P(E)u_n = Q(E) \cdot ae^{\mu h n} ,} \tag{6.22}$$

which is produced by applying time-marching methods to the representative ODE, (4.45). We can express in terms of (6.22) all manner of standard time-marching methods having multiple time steps and various types of intermediate predictor–corrector families. The terms $P(E)$ and $Q(E)$ are polynomials in E referred to as the *characteristic polynomial* and the *particular polynomial*, respectively.

The general solution of (6.22) can be expressed as

$$u_n = \sum_{k=1}^{K} c_k(\sigma_k)^n + ae^{\mu h n} \cdot \frac{Q(e^{\mu h})}{P(e^{\mu h})} , \tag{6.23}$$

where σ_k are the K roots of the characteristic polynomial, $P(\sigma) = 0$. When determinants are involved in the construction of $P(E)$ and $Q(E)$, as would be the case for (6.21), the ratio $Q(E)/P(E)$ can be found by Cramer's rule. Keep in mind that for methods such as (6.21) there are multiple (two in this case) solutions, one for u_n and one for $\tilde{u}_n$, and we are usually only interested in the final solution u_n. Notice also, the important subset of this solution which occurs when $\mu = 0$, representing a time-invariant particular solution, or a steady state. In such a case

$$u_n = \sum_{k=1}^{K} c_k(\sigma_k)^n + a \cdot \frac{Q(1)}{P(1)} .$$

6.4.2 Examples of Solutions to Time-Marching OΔE's

As examples of the use of (6.22) and (6.23), we derive the solutions of (6.19)–(6.21). For the explicit Euler method, (6.19), we have

$$P(E) = E - 1 - \lambda h \qquad (6.24)$$
$$Q(E) = h$$

and the solution of its representative OΔE follows immediately from (6.23):

$$u_n = c_1(1 + \lambda h)^n + ae^{\mu hn} \cdot \frac{h}{e^{\mu h} - 1 - \lambda h} .$$

For the implicit Euler method, (6.20), we have

$$P(E) = (1 - \lambda h)E - 1 \qquad (6.25)$$
$$Q(E) = hE ,$$

so

$$u_n = c_1\left(\frac{1}{1 - \lambda h}\right)^n + ae^{\mu hn} \cdot \frac{he^{\mu h}}{(1 - \lambda h)e^{\mu h} - 1} .$$

In the case of the coupled predictor–corrector equations, (6.21), one solves for the final family u_n (one can also find a solution for the intermediate family $\tilde{u}$), and there results

$$P(E) = \det \begin{bmatrix} E & -(1 + \lambda h) \\ -\frac{1}{2}(1 + \lambda h)E & E - \frac{1}{2} \end{bmatrix} = E\left(E - 1 - \lambda h - \frac{1}{2}\lambda^2 h^2\right)$$

$$Q(E) = \det \begin{bmatrix} E & h \\ -\frac{1}{2}(1 + \lambda h)E & \frac{1}{2}hE \end{bmatrix} = \frac{1}{2}hE(E + 1 + \lambda h) .$$

The σ-root is found from

$$P(\sigma) = \sigma\left(\sigma - 1 - \lambda h - \frac{1}{2}\lambda^2 h^2\right) = 0 ,$$

which has only one nontrivial root ($\sigma = 0$ is simply a shift in the reference index). The complete solution can therefore be written

$$u_n = c_1\left(1 + \lambda h + \frac{1}{2}\lambda^2 h^2\right)^n + ae^{\mu hn} \cdot \frac{\frac{1}{2}h(e^{\mu h} + 1 + \lambda h)}{e^{\mu h} - 1 - \lambda h - \frac{1}{2}\lambda^2 h^2} . \qquad (6.26)$$

6.5 The λ–σ Relation

6.5.1 Establishing the Relation

We have now been introduced to two basic kinds of roots, the λ-roots and the σ-roots. The former are the eigenvalues of the A matrix in the ODE's found by space differencing the original PDE, and the latter are the roots of the characteristic polynomial in a representative OΔE found by applying a time-marching method to the representative ODE. There is a fundamental relation between the two which can be used to identify many of the essential properties of a time-march method. This relation is first demonstrated by developing it for the explicit Euler method.

First we make use of the semi-discrete approach to find a system of ODE's and then express its solution in the form of (4.27). Remembering that $t = nh$, one can write

$$\vec{u}(t) = c_1\left(e^{\lambda_1 h}\right)^n \vec{x}_1 + \cdots + c_m\left(e^{\lambda_m h}\right)^n \vec{x}_m + \cdots + c_M\left(e^{\lambda_M h}\right)^n \vec{x}_M + P.S. \,,$$
(6.27)

where for the present we are not interested in the form of the particular solution ($P.S.$). Now the explicit Euler method produces for each λ-root, one σ-root, which is given by $\sigma = 1 + \lambda h$. So if we use the Euler method for the time advance of the ODE's, the solution[2] of the resulting OΔE is

$$\vec{u}_n = c_1(\sigma_1)^n \vec{x}_1 + \cdots + c_m(\sigma_m)^n \vec{x}_m + \cdots + c_M(\sigma_M)^n \vec{x}_M + P.S. \,,$$
(6.28)

where the c_m and the $\vec{x}_m$ in the two equations are identical and $\sigma_m = (1 + \lambda_m h)$. Comparing (6.27) and (6.28), we see a correspondence between σ_m and $e^{\lambda_m h}$. Since the value of $e^{\lambda h}$ can be expressed in terms of the series

$$e^{\lambda h} = 1 + \lambda h + \frac{1}{2}\lambda^2 h^2 + \frac{1}{6}\lambda^3 h^3 + \cdots + \frac{1}{n!}\lambda^n h^n + \cdots$$

the truncated expansion $\sigma = 1 + \lambda h$ is a reasonable[3] approximation for small enough λh.

Suppose, instead of the Euler method, we use the leapfrog method for the time advance, which is defined by

$$u_{n+1} = u_{n-1} + 2hu'_n \,.$$
(6.29)

Applying (6.29) to (6.8), we have the characteristic polynomial $P(E) = E^2 - 2\lambda h E - 1$, so that for every λ the σ must satisfy the relation

$$\sigma_m^2 - 2\lambda_m h \sigma_m - 1 = 0 \,.$$
(6.30)

---

[2] Based on Section 4.4.
[3] The error is $O(\lambda^2 h^2)$.

Now we notice that each λ produces two σ-roots. For one of these we find

$$\sigma_m = \lambda_m h + \sqrt{1 + \lambda_m^2 h^2} \tag{6.31}$$

$$= 1 + \lambda_m h + \frac{1}{2}\lambda_m^2 h^2 - \frac{1}{8}\lambda_m^4 h^4 + \cdots . \tag{6.32}$$

This is an approximation to $e^{\lambda_m h}$ with an error $O(\lambda^3 h^3)$. The other root, $\lambda_m h - \sqrt{1 + \lambda_m^2 h^2}$, will be discussed in Section 6.5.3.

6.5.2 The Principal σ-Root

Based on the above we make the following observation:

> Application of the same time-marching method to all of the equations in a coupled system of linear ODE's in the form of (4.6) always produces one σ-root for every λ-root that satisfies the relation
>
> $$\sigma = 1 + \lambda h + \frac{1}{2}\lambda^2 h^2 + \cdots + \frac{1}{k!}\lambda^k h^k + O(h^{k+1}) ,$$
>
> where k is the order of the time-marching method.

(6.33)

We refer to the root that has the above property as the *principal σ-root*, and designate it $(\sigma_m)_1$. The above property can be stated regardless of the details of the time-marching method, knowing only that its leading error is $O(h^{k+1})$. Thus the principal root is an approximation to $e^{\lambda h}$ up to $O(h^k)$.

Note that a second-order approximation to a derivative written in the form

$$(\delta_t u)_n = \frac{1}{2h}(u_{n+1} - u_{n-1}) \tag{6.34}$$

has a leading truncation error which is $O(h^2)$, while the second-order time-marching method which results from this approximation, which is the leapfrog method:

$$u_{n+1} = u_{n-1} + 2h u_n' \tag{6.35}$$

has a leading truncation error $O(h^3)$. This arises simply because of our notation for the time-marching method in which we have multiplied through by h to get an approximation for the *function* u_{n+1} rather than the *derivative* as in (6.34). The following example makes this clear. Consider a solution obtained at a given time T using a second-order time-marching method with a time step h. Now consider the solution obtained using the same method with a time step $h/2$. Since the error per time step is $O(h^3)$, this is reduced by a factor of eight (considering the leading term only). However, twice as many time steps are required to reach the time T. Therefore the error at the end of the simulation is reduced by a factor of four, consistent with a second-order approximation.

6.5.3 Spurious σ-Roots

We saw from (6.30) that the λ–σ relation for the leapfrog method produces two σ-roots for each λ. One of these we identified as the principal root, which always has the property given in (6.33). The other is referred to as a *spurious* σ-root and designated $(\sigma_m)_2$. In general, the λ–σ relation produced by a time-marching scheme can result in multiple σ-roots all of which, except for the principal one, are spurious. All spurious roots are designated $(\sigma_m)_k$ where $k = 2, 3, \cdots$. No matter whether a σ-root is principal or spurious, it is always some algebraic function of the product λh. To express this fact we use the notation $\sigma = \sigma(\lambda h)$.

If a time-marching method produces spurious σ-roots, the solution for the OΔE in the form shown in (6.28) must be modified. Following again the message of Section 4.4, we have

$$
\begin{aligned}
\vec{u}_n = \; &c_{11}(\sigma_1)_1^n \, \vec{x}_1 + \cdots + c_{m1}(\sigma_m)_1^n \, \vec{x}_m + \cdots + c_{M1}(\sigma_M)_1^n \, \vec{x}_M + P.S. \\
&+ c_{12}(\sigma_1)_2^n \, \vec{x}_1 + \cdots + c_{m2}(\sigma_m)_2^n \, \vec{x}_m + \cdots + c_{M2}(\sigma_M)_2^n \, \vec{x}_M \\
&+ c_{13}(\sigma_1)_3^n \, \vec{x}_1 + \cdots + c_{m3}(\sigma_m)_3^n \, \vec{x}_m + \cdots + c_{M3}(\sigma_M)_3^n \, \vec{x}_M \\
&+ \text{etc., if there are more spurious roots.} \hspace{2cm} (6.36)
\end{aligned}
$$

Spurious roots arise if a method uses data from time level $n-1$ or earlier to advance the solution from time level n to $n+1$. Such roots originate entirely from the numerical approximation of the time-marching method and have nothing to do with the ODE being solved. However, generation of spurious roots does not, in itself, make a method inferior. In fact, many very accurate methods in practical use for integrating some forms of ODE's have spurious roots.

It should be mentioned that methods with spurious roots are not self-starting. For example, if there is one spurious root to a method, all of the coefficients $(c_m)_2$ in (6.36) must be initialized by some starting procedure. The initial vector $\vec{u}_0$ does not provide enough data to initialize all of the coefficients. This results because methods which produce spurious roots require data from time level $n-1$ or earlier. For example, the leapfrog method requires $\vec{u}_{n-1}$ and thus cannot be started using only $\vec{u}_n$.

Presumably (i.e. if one starts the method properly[4]) the spurious coefficients are all initialized with very small magnitudes, and presumably the magnitudes of the spurious roots themselves are all less than one (see Chapter 7). Then the presence of spurious roots does not contaminate the answer. That is, after some finite time, the amplitude of the error associated with the spurious roots is even smaller then when it was initialized. Thus while spurious roots must be considered in *stability* analysis, they play virtually no role in *accuracy* analysis.

---

[4] Normally using a self-starting method for the first step or steps, as required.

6.5.4 One-Root Time-Marching Methods

There are a number of time-marching methods that produce only one σ-root for each λ-root. We refer to them as *one-root* methods. They are also called one-step methods. They have the significant advantage of being self-starting which carries with it the very useful property that the time-step interval can be changed at will throughout the marching process. Three one-root methods were analyzed in Section 6.4.2. A popular method having this property, the so-called θ-method, is given by the formula

$$u_{n+1} = u_n + h\big[(1-\theta)u_n' + \theta u_{n+1}'\big] .$$

The θ-method represents the explicit Euler $(\theta = 0)$, the trapezoidal $(\theta = \frac{1}{2})$, and the implicit Euler methods $(\theta = 1)$, respectively. Its λ–σ relation is

$$\sigma = \frac{1+(1-\theta)\lambda h}{1-\theta\lambda h} .$$

It is instructive to compare the exact solution to a set of ODE's (with a complete eigensystem) having time-invariant forcing terms with the exact solution to the OΔE's for one-root methods. These are

$$\vec{u}(t) = c_1\big(e^{\lambda_1 h}\big)^n \vec{x}_1 + \cdots + c_m\big(e^{\lambda_m h}\big)^n \vec{x}_m + \cdots + c_M\big(e^{\lambda_M h}\big)^n \vec{x}_M + A^{-1}\vec{f}$$
$$\vec{u}_n = c_1(\sigma_1)^n \vec{x}_1 + \cdots + c_m(\sigma_m)^n \vec{x}_m + \cdots + c_M(\sigma_M)^n \vec{x}_M + A^{-1}\vec{f} ,$$
$$(6.37)$$

respectively. Notice that when $t = 0$ and $n = 0$, these equations are identical, so that all the constants, vectors, and matrices are identical except the $\vec{u}$ and the terms inside the parentheses on the right hand sides. The only error made by introducing the time marching is the error that σ makes in approximating $e^{\lambda h}$.

6.6 Accuracy Measures of Time-Marching Methods

6.6.1 Local and Global Error Measures

There are two broad categories of errors that can be used to derive and evaluate time-marching methods. One is the error made in each time step. This is a *local* error such as that found from a Taylor table analysis (see Section 3.4). It is usually used as the basis for establishing the order of a method. The other is the error determined at the end of a given *event* which has covered a specific interval of time composed of many time steps. This is a *global* error. It is useful for comparing methods, as we shall see in Chapter 8.

It is quite common to judge a time-marching method on the basis of results found from a Taylor table. However, a Taylor series analysis is a very limited tool for finding the more subtle properties of a numerical time-marching method. For example, it is of no use in

- finding spurious roots,
- evaluating numerical stability and separating the errors in phase and amplitude,
- analyzing the particular solution of predictor–corrector combinations,
- finding the global error.

The latter three of these are of concern to us here, and to study them we make use of the material developed in the previous sections of this chapter. Our error measures are based on the difference between the exact solution to the representative ODE, given by

$$u(t) = c\,e^{\lambda t} + \frac{ae^{\mu t}}{\mu - \lambda} \tag{6.38}$$

and the solution to the representative $O\Delta E$'s, including only the contribution from the principal root, which can be written as

$$u_n = c_1(\sigma_1)^n + ae^{\mu hn} \cdot \frac{Q(e^{\mu h})}{P(e^{\mu h})} \ . \tag{6.39}$$

6.6.2 Local Accuracy of the Transient Solution $(er_\lambda, |\sigma|, er_\omega)$

Transient Error. The particular choice of an error measure, either local or global, is to some extent arbitrary. However, a necessary condition for the choice should be that the measure can be used consistently for all methods. In the discussion of the λ–σ relation we saw that all time-marching methods produce a principal σ-root for every λ-root that exists in a set of linear ODE's. Therefore, a very natural local error measure for the transient solution is the value of the difference between solutions based on these two roots. We designate this by er_λ and make the following definition

$$\boxed{er_\lambda \equiv e^{\lambda h} - \sigma_1 \ .}$$

The leading error term can be found by expanding in a Taylor series and choosing the first nonvanishing term. This is similar to the error found from a Taylor table. The *order* of the method is the last power of λh matched exactly.

Amplitude and Phase Error. Suppose a λ eigenvalue is imaginary. Such can indeed be the case when we study the equations governing periodic convection, which produces harmonic motion. For such cases it is more meaningful to express the error in terms of amplitude and phase. Let $\lambda = i\omega$ where ω

is a real number representing a frequency. Then the numerical method must produce a principal σ-root that is complex and expressible in the form

$$\sigma_1 = \sigma_{\text{real}} + i\sigma_{\text{imaginary}} \approx e^{i\omega h} . \tag{6.40}$$

From this it follows that the local error in amplitude is measured by the deviation of $|\sigma_1|$ from unity, that is

$$\boxed{er_{\text{a}} = 1 - |\sigma_1| = 1 - \sqrt{(\sigma_1)^2_{\text{real}} + (\sigma_1)^2_{\text{imaginary}}}}$$

and the local error in phase can be defined as

$$\boxed{er_\omega \equiv \omega h - \tan^{-1}\left[(\sigma_1)_{\text{imaginary}}/(\sigma_1)_{\text{real}}\right] .} \tag{6.41}$$

Amplitude and phase errors are important measures of the suitability of time-marching methods for convection and wave propagation phenomena.

The approach to error analysis described in Section 3.5 can be extended to the combination of a spatial discretization and a time-marching method applied to the linear convection equation. The principal root, $\sigma_1(\lambda h)$, is found using $\lambda = -ia\kappa^*$, where κ^* is the modified wavenumber of the spatial discretization. Introducing the Courant number, $C_n = ah/\Delta x$, we have $\lambda h = -iC_n\kappa^*\Delta x$. Thus one can obtain values of the principal root over the range $0 \le \kappa\Delta x \le \pi$ for a given value of the Courant number. The above expression for er_ω can be normalized to give the error in the phase speed, as follows

$$er_{\text{p}} = \frac{er_\omega}{\omega h} = 1 + \frac{\tan^{-1}\left[(\sigma_1)_{\text{imaginary}}/(\sigma_1)_{\text{real}}\right]}{C_n\kappa\Delta x} \tag{6.42}$$

where $\omega = -a\kappa$. A positive value of er_{p} corresponds to phase lag (the numerical phase speed is too small), while a negative value corresponds to phase lead (the numerical phase speed is too large).

6.6.3 Local Accuracy of the Particular Solution (er_μ)

The numerical error in the particular solution is found by comparing the particular solution of the ODE with that for the OΔE. We have found these to be given by

$$P.S._{\text{(ODE)}} = ae^{\mu t} \cdot \frac{1}{(\mu - \lambda)}$$

and

$$P.S._{\text{(O}\Delta\text{E)}} = ae^{\mu t} \cdot \frac{Q(e^{\mu h})}{P(e^{\mu h})} ,$$

respectively. For a measure of the *local* error in the particular solution we introduce the definition

$$er_\mu \equiv h\left\{\frac{P.S._{(O\Delta E)}}{P.S._{(ODE)}} - 1\right\}. \tag{6.43}$$

The multiplication by h converts the error from a global measure to a local one, so that the order of er_λ and er_μ are consistent. In order to determine the leading error term, (6.43) can be written in terms of the characteristic and particular polynomials as

$$\boxed{er_\mu = \frac{c_0}{\mu - \lambda} \cdot \left[(\mu - \lambda)Q(e^{\mu h}) - P(e^{\mu h})\right]} \tag{6.44}$$

and expanded in a Taylor series, where

$$c_0 = \lim_{h \to 0} \frac{h(\mu - \lambda)}{P(e^{\mu h})}.$$

The value of c_0 is a method-dependent constant that is often equal to one. If the forcing function is independent of time, μ is equal to zero, and for this case, many numerical methods generate an er_μ that is also zero.

The algebra involved in finding the order of er_μ can be quite tedious. However, this order is quite important in determining the true order of a time-marching method by the process that has been outlined. An illustration of this is given in the section on Runge–Kutta methods.

6.6.4 Time Accuracy for Nonlinear Applications

In practice, time-marching methods are usually applied to nonlinear ODE's, and it is necessary that the advertised order of accuracy be valid for the nonlinear cases as well as for the linear ones. A necessary condition for this to occur is that the local accuracies of *both the transient and the particular solutions be of the same order*. More precisely, a time-marching method is said to be of order k if

$$er_\lambda = c_1 \cdot h^{k_1+1} \tag{6.45}$$
$$er_\mu = c_2 \cdot h^{k_2+1}, \tag{6.46}$$

where

$$k = \text{smallest of } (k_1, k_2). \tag{6.47}$$

The reader should be aware that this is not sufficient. For example, to derive all of the necessary conditions for the fourth-order Runge–Kutta method presented later in this chapter the derivation must be performed for a nonlinear ODE. However, the analysis based on a linear nonhomogeneous ODE produces the appropriate conditions for the majority of time-marching methods used in CFD.

6.6.5 Global Accuracy

In contrast to the local error measures which have just been discussed, we can also define global error measures. These are useful when we come to the evaluation of time-marching methods for specific purposes. This subject is covered in Chapter 8 after our introduction to stability in Chapter 7.

Suppose we wish to compute some time-accurate phenomenon over a fixed interval of time using a constant time step. We refer to such a computation as an "event". Let T be the fixed time of the event and h be the chosen step size. Then the required number of time steps, N, is given by the relation

$$T = Nh .$$

Global Error in the Transient. A natural extension of er_λ to cover the error in an entire event is given by

$$Er_\lambda \equiv e^{\lambda T} - [\sigma_1(\lambda h)]^N . \tag{6.48}$$

Global Error in Amplitude and Phase. If the event is periodic, we are more concerned with the global error in amplitude and phase. These are given by

$$Er_a = 1 - \left[\sqrt{(\sigma_1)^2_{\text{real}} + (\sigma_1)^2_{\text{imaginary}}} \right]^N \tag{6.49}$$

and

$$Er_\omega \equiv N\left[\omega h - \tan^{-1}\left(\frac{(\sigma_1)_{\text{imaginary}}}{(\sigma_1)_{\text{real}}} \right) \right]$$
$$= \omega T - N \tan^{-1}\left[(\sigma_1)_{\text{imaginary}}/(\sigma_1)_{\text{real}} \right] . \tag{6.50}$$

Global Error in the Particular Solution. Finally, the global error in the particular solution follows naturally by comparing the solutions to the ODE and the OΔE. It can be measured by

$$Er_\mu \equiv (\mu - \lambda)\frac{Q(e^{\mu h})}{P(e^{\mu h})} - 1 .$$

6.7 Linear Multistep Methods

In the previous sections, we have developed the framework of error analysis for time advance methods and have randomly introduced a few methods without addressing motivational, developmental, or design issues. In the subsequent sections, we introduce classes of methods along with their associated error analysis. We shall not spend much time on development or design of these methods, since most of them have historic origins from a wide variety of disciplines and applications. The linear multistep methods (LMM's) are probably the most natural extension to time marching of the space differencing schemes introduced in Chapter 3 and can be analyzed for accuracy or designed using the Taylor table approach of Section 3.4.

6.7.1 The General Formulation

When applied to the nonlinear ODE

$$\frac{du}{dt} = u' = F(u,t) ,$$

all linear multistep methods can be expressed in the general form

$$\sum_{k=1-K}^{1} \alpha_k u_{n+k} = h \sum_{k=1-K}^{1} \beta_k F_{n+k} , \qquad (6.51)$$

where the notation for F is defined in Section 6.1. The methods are said to be linear because the α's and β's are independent of u and n, and they are said to be K-step because K time-levels of data are required to march the solution one time-step, h. They are *explicit* if $\beta_1 = 0$ and *implicit* otherwise.

When (6.51) is applied to the representative equation, (6.8), and the result is expressed in operational form, one finds

$$\left(\sum_{k=1-K}^{1} \alpha_k E^k \right) u_n = h \left(\sum_{k=1-K}^{1} \beta_k E^k \right) (\lambda u_n + a e^{\mu h n}) . \qquad (6.52)$$

We recall from Section 6.5.2 that a time-marching method when applied to the representative equation must provide a σ-root, labeled σ_1, that approximates $e^{\lambda h}$ through the order of the method. The condition referred to as *consistency* simply means that $\sigma \to 1$ as $h \to 0$, and it is certainly a necessary condition for the accuracy of any time-marching method. We can also agree that, to be of any value in time accuracy, a method should at least be first-order accurate, that is $\sigma \to (1 + \lambda h)$ as $h \to 0$. One can show that these conditions are met by any method represented by (6.51) if

$$\sum_{k} \alpha_k = 0 \quad \text{and} \quad \sum_{k} \beta_k = \sum_{k} (K + k - 1) \alpha_k .$$

Since both sides of (6.51) can be multiplied by an arbitrary constant, these methods are often "normalized" by requiring

$$\sum_{k} \beta_k = 1 .$$

Under this condition, $c_0 = 1$ in (6.44).

6.7.2 Examples

There are many special explicit and implicit forms of linear multistep methods. Two well-known families of them, referred to as Adams–Bashforth (*explicit*) and Adams–Moulton (*implicit*), can be designed using the Taylor table approach of Section 3.4. The Adams–Moulton family is obtained from (6.51) with

$$\alpha_1 = 1, \qquad \alpha_0 = -1, \qquad \alpha_k = 0, \qquad k = -1, -2, \cdots . \qquad (6.53)$$

The Adams–Bashforth family has the same α's with the additional constraint that $\beta_1 = 0$. The three-step Adams–Moulton method can be written in the following form

$$u_{n+1} = u_n + h(\beta_1 u'_{n+1} + \beta_0 u'_n + \beta_{-1} u'_{n-1} + \beta_{-2} u'_{n-2}) . \qquad (6.54)$$

A Taylor table for (6.54) can be generated as shown in Table 6.1.

This leads to the linear system

$$
\begin{bmatrix}
1 & 1 & 1 & 1 \\
2 & 0 & -2 & -4 \\
3 & 0 & 3 & 12 \\
4 & 0 & -4 & -32
\end{bmatrix}
\begin{bmatrix}
\beta_1 \\
\beta_0 \\
\beta_{-1} \\
\beta_{-2}
\end{bmatrix}
=
\begin{bmatrix}
1 \\
1 \\
1 \\
1
\end{bmatrix}
\qquad (6.55)
$$

to solve for the β's, resulting in

$$\beta_1 = 9/24, \qquad \beta_0 = 19/24, \qquad \beta_{-1} = -5/24, \qquad \beta_{-2} = 1/24 . \qquad (6.56)$$

which produces a method which is fourth-order accurate.[5] With $\beta_1 = 0$ one obtains

$$
\begin{bmatrix}
1 & 1 & 1 \\
0 & -2 & -4 \\
0 & 3 & 12
\end{bmatrix}
\begin{bmatrix}
\beta_0 \\
\beta_{-1} \\
\beta_{-2}
\end{bmatrix}
=
\begin{bmatrix}
1 \\
1 \\
1
\end{bmatrix}
\qquad (6.57)
$$

giving

$$\beta_0 = 23/12, \qquad \beta_{-1} = -16/12, \qquad \beta_{-2} = 5/12 . \qquad (6.58)$$

This is the third-order Adams–Bashforth method.

A list of simple methods, some of which are very common in CFD applications, is given below together with identifying names that are sometimes associated with them. In the following material ABn and AMn are used as abbreviations for the nth-order Adams–Bashforth and nth-order Adams–Moulton methods. One can verify that the Adams type schemes given below satisfy (6.55) and (6.57) up to the order of the method.

Explicit Methods.

$$
\begin{array}{lll}
u_{n+1} & = u_n + hu'_n & \text{Euler} \\
u_{n+1} & = u_{n-1} + 2hu'_n & \text{Leapfrog} \\
u_{n+1} & = u_n + \frac{1}{2}h\left[3u'_n - u'_{n-1}\right] & \text{AB2} \\
u_{n+1} & = u_n + \frac{h}{12}\left[23u'_n - 16u'_{n-1} + 5u'_{n-2}\right] & \text{AB3}
\end{array}
$$

---

[5] Recall from Section 6.5.2 that a kth-order time-marching method has a leading truncation error term which is $O(h^{k+1})$.

Table 6.1. Taylor table for the Adams–Moulton three-step linear multistep method

	u_n	$h\cdot u'_n$	$h^2\cdot u''_n$	$h^3\cdot u'''_n$	$h^4\cdot u''''_n$
		1	$\frac{1}{2}$	$\frac{1}{6}$	$\frac{1}{24}$
u_{n+1}	1				
$-u_n$	-1				
$-h\beta_1 u'_{n+1}$		$-\beta_1$	$-\beta_1$	$-\beta_1\frac{1}{2}$	$-\beta_1\frac{1}{6}$
$-h\beta_0 u'_n$		$-\beta_0$			
$-h\beta_{-1}u'_{n-1}$		$-\beta_{-1}$	β_{-1}	$-\beta_{-1}\frac{1}{2}$	$\beta_{-1}\frac{1}{6}$
$-h\beta_{-2}u'_{n-2}$		$-(-2)^0\beta_{-2}$	$-(-2)^1\beta_{-2}$	$-(-2)^2\beta_{-2}\frac{1}{2}$	$-(-2)^3\beta_{-2}\frac{1}{6}$

Implicit Methods.

$$u_{n+1} = u_n + hu'_{n+1} \qquad\qquad \text{Implicit Euler}$$
$$u_{n+1} = u_n + \tfrac{1}{2}h[u'_n + u'_{n+1}] \qquad\qquad \text{Trapezoidal (AM2)}$$
$$u_{n+1} = \tfrac{1}{3}[4u_n - u_{n-1} + 2hu'_{n+1}] \qquad\qquad \text{2nd-order Backward}$$
$$u_{n+1} = u_n + \tfrac{h}{12}[5u'_{n+1} + 8u'_n - u'_{n-1}] \qquad\qquad \text{AM3}$$

6.7.3 Two-Step Linear Multistep Methods

High-resolution CFD problems usually require very large data sets to store the spatial information from which the time derivative is calculated. This limits the interest in multistep methods to about two time levels. The most general two-step linear multistep method (i.e. $K = 2$ in (6.51)) that is at least first-order accurate can be written as

$$(1 + \xi)u_{n+1} = [(1 + 2\xi)u_n - \xi u_{n-1}] + h[\,\theta u'_{n+1} + (1 - \theta + \varphi)u'_n - \varphi u'_{n-1}]\,. \tag{6.59}$$

Clearly the methods are explicit if $\theta = 0$ and implicit otherwise. A list of methods contained in (6.59) is given in Table 6.2. Notice that the Adams methods have $\xi = 0$, which corresponds to $\alpha_{-1} = 0$ in (6.51). Methods with $\xi = -1/2$, which corresponds to $\alpha_0 = 0$ in (6.51), are known as *Milne* methods.

Table 6.2. Some linear one- and two-step methods; see (6.59)

θ	ξ	φ	Method	Order
0	0	0	Euler	1
1	0	0	Implicit Euler	1
1/2	0	0	Trapezoidal or AM2	2
1	1/2	0	2nd-order Backward	2
3/4	0	-1/4	Adams type	2
1/3	-1/2	-1/3	Lees	2
1/2	-1/2	-1/2	Two-step trapezoidal	2
5/9	-1/6	-2/9	A-contractive	2
0	-1/2	0	Leapfrog	2
0	0	1/2	AB2	2
0	-5/6	-1/3	Most accurate explicit	3
1/3	-1/6	0	Third-order implicit	3
5/12	0	1/12	AM3	3
1/6	-1/2	-1/6	Milne	4

One can show after a little algebra that both er_μ and er_λ are reduced to $O(h^3)$ (i.e. the methods are 2nd-order accurate) if

$$\varphi = \xi - \theta + \frac{1}{2} \ .$$

The class of all 3rd-order methods is determined by imposing the additional constraint

$$\xi = 2\theta - \frac{5}{6} \ .$$

Finally a unique 4th-order method is found by setting $\theta = -\varphi = -\xi/3 = \frac{1}{6}$.

6.8 Predictor–Corrector Methods

There are a wide variety of predictor–corrector schemes created and used for a variety of purposes. Their use in solving ODE's is relatively easy to illustrate and understand. Their use in solving PDE's can be much more subtle and demands concepts[6] which have no counterpart in the analysis of ODE's.

Predictor–corrector methods constructed to time-march linear or nonlinear ODE's are composed of sequences of linear multistep methods, each of which is referred to as a family in the solution process. There may be many families in the sequence, and usually the final family has a higher Taylor-series order of accuracy than the intermediate ones. Their use is motivated by ease of application and increased efficiency, where measures of efficiency are discussed in the next two chapters.

A simple one-predictor–one-corrector example is given by

$$\tilde{u}_{n+\alpha} = u_n + \alpha h u'_n$$
$$u_{n+1} = u_n + h\left[\beta \tilde{u}'_{n+\alpha} + \gamma u'_n\right] , \qquad (6.60)$$

where the parameters α, β and γ are arbitrary parameters to be determined. One can analyze this sequence by applying it to the representative equation and using the operational techniques outlined in Section 6.4. It is easy to show, following the example leading to (6.26), that

$$P(E) = E^\alpha \cdot \left[E - 1 - (\gamma + \beta)\lambda h - \alpha\beta\lambda^2 h^2\right] \qquad (6.61)$$
$$Q(E) = E^\alpha \cdot h \cdot (\beta E^\alpha + \gamma + \alpha\beta\lambda h) \ . \qquad (6.62)$$

Considering only local accuracy, one is led, by following the discussion in Section 6.6, to the following observations. For the method to be second-order accurate *both* er_λ and er_μ must be $O(h^3)$. For this to hold for er_λ, it is obvious from (6.61) that

---

[6] Such as alternating direction, fractional-step, and hybrid methods.

$$\gamma + \beta = 1 \quad ; \quad \alpha\beta = \frac{1}{2} \,,$$

which provides two equations for three unknowns. The situation for er_μ requires some algebra, but it is not difficult to show using (6.44) that the same conditions also make it $O(h^3)$. One concludes, therefore, that the predictor–corrector sequence

$$\tilde{u}_{n+\alpha} = u_n + \alpha h u'_n$$
$$u_{n+1} = u_n + \frac{1}{2}h\left[\left(\frac{1}{\alpha}\right)\tilde{u}'_{n+\alpha} + \left(\frac{2\alpha - 1}{\alpha}\right)u'_n\right] \qquad (6.63)$$

is a second-order accurate method for any α.

A classical predictor–corrector sequence is formed by following an Adams–Bashforth predictor of any order with an Adams–Moulton corrector having an order one higher. The order of the combination is then equal to the order of the corrector. If the order of the corrector is k, we refer to these as ABM(k) methods. The Adams–Bashforth–Moulton sequence for $k = 3$ is

$$\tilde{u}_{n+1} = u_n + \frac{1}{2}h\left(3u'_n - u'_{n-1}\right)$$
$$u_{n+1} = u_n + \frac{h}{12}\left(5\tilde{u}'_{n+1} + 8u'_n - u'_{n-1}\right) \,. \qquad (6.64)$$

Some simple, specific, second-order accurate methods are given below. The Gazdag method, which we discuss in Chapter 8, is

$$\tilde{u}_{n+1} = u_n + \frac{1}{2}h\left(3\tilde{u}'_n - \tilde{u}'_{n-1}\right)$$
$$u_{n+1} = u_n + \frac{1}{2}h\left(\tilde{u}'_n + \tilde{u}'_{n+1}\right) \,. \qquad (6.65)$$

The Burstein method, obtained from (6.63) with $\alpha = 1/2$ is

$$\tilde{u}_{n+1/2} = u_n + \frac{1}{2}hu'_n$$
$$u_{n+1} = u_n + h\tilde{u}'_{n+1/2} \,. \qquad (6.66)$$

and, finally, MacCormack's method, presented earlier in this chapter, is

$$\tilde{u}_{n+1} = u_n + hu'_n$$
$$u_{n+1} = \frac{1}{2}\left(u_n + \tilde{u}_{n+1} + h\tilde{u}'_{n+1}\right) \,. \qquad (6.67)$$

Note that MacCormack's method can also be written as

$$\tilde{u}_{n+1} = u_n + hu'_n$$
$$u_{n+1} = u_n + \frac{1}{2}h(u'_n + \tilde{u}'_{n+1}) \,. \qquad (6.68)$$

from which it is clear that it is obtained from (6.63) with $\alpha = 1$.

6.9 Runge–Kutta Methods

There is a special subset of predictor–corrector methods, referred to as Runge–Kutta methods,[7] that produce just one σ-root for each λ-root such that $\sigma(\lambda h)$ corresponds to the Taylor series expansion of $e^{\lambda h}$ out through the order of the method and then truncates. Thus for a Runge–Kutta method of order k (up to 4th order), the principal (and only) σ-root is given by

$$\sigma = 1 + \lambda h + \frac{1}{2}\lambda^2 h^2 + \cdots + \frac{1}{k!}\lambda^k h^k . \tag{6.69}$$

It is not particularly difficult to build this property into a method, but, as we pointed out in Section 6.6.4, it is not sufficient to guarantee kth-order accuracy for the solution of $u' = F(u, t)$ or for the representative equation. To ensure kth-order accuracy, the method must further satisfy the constraint that

$$er_\mu = O(h^{k+1}) \tag{6.70}$$

and this is much more difficult.

The most widely publicized Runge–Kutta process is the one that leads to the fourth-order method. We present it below in some detail. It is usually introduced in the form

$$k_1 = hF(u_n, t_n)$$
$$k_2 = hF(u_n + \beta k_1, t_n + \alpha h)$$
$$k_3 = hF(u_n + \beta_1 k_1 + \gamma_1 k_2, t_n + \alpha_1 h)$$
$$k_4 = hF(u_n + \beta_2 k_1 + \gamma_2 k_2 + \delta_2 k_3, t_n + \alpha_2 h)$$

followed by

$$u(t_n + h) - u(t_n) = \mu_1 k_1 + \mu_2 k_2 + \mu_3 k_3 + \mu_4 k_4 . \tag{6.71}$$

However, we prefer to present it using predictor-corrector notation. Thus, a scheme entirely equivalent to (6.71) is

$$\widehat{u}_{n+\alpha} = u_n + \beta h u'_n$$
$$\tilde{u}_{n+\alpha_1} = u_n + \beta_1 h u'_n + \gamma_1 h \widehat{u}'_{n+\alpha}$$
$$\overline{u}_{n+\alpha_2} = u_n + \beta_2 h u'_n + \gamma_2 h \widehat{u}'_{n+\alpha} + \delta_2 h \tilde{u}'_{n+\alpha_1}$$
$$u_{n+1} = u_n + \mu_1 h u'_n + \mu_2 h \widehat{u}'_{n+\alpha} + \mu_3 h \tilde{u}'_{n+\alpha_1} + \mu_4 h \overline{u}'_{n+\alpha_2} . \tag{6.72}$$

Appearing in (6.71) and (6.72) are a total of 13 parameters which are to be determined such that the method is fourth-order according to the requirements in (6.69) and (6.70). First of all, the choices for the time samplings, α, α_1, and α_2, are not arbitrary. They must satisfy the relations

---

[7] Although implicit and multi-step Runge–Kutta methods exist, we will consider only single-step, explicit Runge–Kutta methods here.

$$\alpha = \beta$$
$$\alpha_1 = \beta_1 + \gamma_1 \tag{6.73}$$
$$\alpha_2 = \beta_2 + \gamma_2 + \delta_2 .$$

The algebra involved in finding algebraic equations for the remaining 10 parameters is not trivial, but the equations follow directly from finding $P(E)$ and $Q(E)$ and then satisfying the conditions in (6.69) and (6.70). Using (6.73) to eliminate the β's we find from (6.69) the four conditions

$$
\begin{aligned}
\mu_1 + \mu_2 + \mu_3 + \mu_4 &= 1 &\quad(1)\\
\mu_2\alpha + \mu_3\alpha_1 + \mu_4\alpha_2 &= 1/2 &\quad(2)\\
\mu_3\alpha\gamma_1 + \mu_4(\alpha\gamma_2 + \alpha_1\delta_2) &= 1/6 &\quad(3)\\
\mu_4\alpha\gamma_1\delta_2 &= 1/24 &\quad(4) .
\end{aligned}
\tag{6.74}
$$

These four relations guarantee that the five terms in σ exactly match the first five terms in the expansion of $e^{\lambda h}$. To satisfy the condition that $er_\mu = O(h^5)$, we have to fulfill four more conditions

$$
\begin{aligned}
\mu_2\alpha^2 + \mu_3\alpha_1^2 + \mu_4\alpha_2^2 &= 1/3 &\quad(3)\\
\mu_2\alpha^3 + \mu_3\alpha_1^3 + \mu_4\alpha_2^3 &= 1/4 &\quad(4)\\
\mu_3\alpha^2\gamma_1 + \mu_4(\alpha^2\gamma_2 + \alpha_1^2\delta_2) &= 1/12 &\quad(4)\\
\mu_3\alpha\alpha_1\gamma_1 + \mu_4\alpha_2(\alpha\gamma_2 + \alpha_1\delta_2) &= 1/8 &\quad(4) .
\end{aligned}
\tag{6.75}
$$

The number in parentheses at the end of each equation indicates the order that is the basis for the equation. Thus if the first three equations in (6.74) and the first equation in (6.75) are all satisfied, the resulting method would be third-order accurate. As discussed in Section 6.6.4, the fourth condition in (6.75) cannot be derived using the methodology presented here, which is based on a linear nonhomogenous representative ODE. A more general derivation based on a nonlinear ODE can be found in several books.[8]

There are eight equations in (6.74) and (6.75) which must be satisfied by the ten unknowns. Since the equations are underdetermined, two parameters can be set arbitrarily. Several choices for the parameters have been proposed, but the most popular one is due to Runge. It results in the "standard" fourth-order Runge–Kutta method expressed in predictor–corrector form as

$$\widehat{u}_{n+1/2} = u_n + \frac{1}{2}hu'_n$$

$$\tilde{u}_{n+1/2} = u_n + \frac{1}{2}h\widehat{u}'_{n+1/2}$$

$$\overline{u}_{n+1} = u_n + h\tilde{u}'_{n+1/2}$$

$$u_{n+1} = u_n + \frac{1}{6}h\left[u'_n + 2\left(\widehat{u}'_{n+1/2} + \tilde{u}'_{n+1/2}\right) + \overline{u}'_{n+1}\right] . \tag{6.76}$$

---

[8] The present approach based on a linear nonhomogeneous equation provides all of the necessary conditions for Runge–Kutta methods of up to third order.

Notice that this represents the simple sequence of conventional linear multi-step methods referred to, respectively, as

$$\left.\begin{array}{l} \text{Euler Predictor} \\ \text{Euler Corrector} \\ \text{Leapfrog Predictor} \\ \text{Milne Corrector} \end{array}\right\} \equiv RK4 \ .$$

One can easily show that both the Burstein and the MacCormack methods given by (6.66) and (6.67) are second-order Runge–Kutta methods, and third-order methods can be derived from (6.72) by setting $\mu_4 = 0$ and satisfying only the first three equations in (6.74) and the first equation in (6.75). It is clear that for orders one through four, RK methods of order k require k evaluations of the derivative function to advance the solution one time step. We shall discuss the consequences of this in Chapter 8. Higher-order Runge–Kutta methods can be developed, but they require more derivative evaluations than their order. For example, a fifth-order method requires six evaluations to advance the solution one step. In any event, storage requirements reduce the usefulness of Runge–Kutta methods of order higher than four for CFD applications.

6.10 Implementation of Implicit Methods

We have presented a wide variety of time-marching methods and shown how to derive their λ–σ relations. In the next chapter, we will see that these methods can have widely different properties with respect to stability. This leads to various trade-offs which must be considered in selecting a method for a specific application. Our presentation of the time-marching methods in the context of a linear scalar equation obscures some of the issues involved in implementing an implicit method for systems of equations and nonlinear equations. These are covered in this section.

6.10.1 Application to Systems of Equations

Consider first the numerical solution of our representative ODE

$$u' = \lambda u + ae^{\mu t} \tag{6.77}$$

using the implicit Euler method. Following the steps outlined in Section 6.2, we obtained

$$(1 - \lambda h)u_{n+1} - u_n = he^{\mu h} \cdot ae^{\mu h n} \ . \tag{6.78}$$

Solving for u_{n+1} gives

$$u_{n+1} = \frac{1}{1 - \lambda h}(u_n + h e^{\mu h} \cdot a e^{\mu h n}) . \tag{6.79}$$

This calculation does not seem particularly onerous in comparison with the application of an explicit method to this ODE, requiring only an additional division.

Now let us apply the implicit Euler method to our generic system of equations given by

$$\vec{u}' = A\vec{u} - \vec{f}(t) , \tag{6.80}$$

where $\vec{u}$ and $\vec{f}$ are vectors and we still assume that A is not a function of $\vec{u}$ or t. Now the equivalent to (6.78) is

$$(I - hA)\vec{u}_{n+1} - \vec{u}_n = -h\vec{f}(t + h) \tag{6.81}$$

or

$$\vec{u}_{n+1} = (I - hA)^{-1}[\vec{u}_n - h\vec{f}(t + h)] . \tag{6.82}$$

The inverse is not actually performed, but rather we solve (6.81) as a linear system of equations. For our one-dimensional examples, the system of equations which must be solved is tridiagonal (e.g. for biconvection, $A = -aB_p(-1, 0, 1)/2\Delta x$), and hence its solution is inexpensive, but in multidimensions the bandwidth can be very large. In general, the cost per time step of an implicit method is larger than that of an explicit method. The primary area of application of implicit methods is in the solution of *stiff* ODE's, as we shall see in Chapter 8.

6.10.2 Application to Nonlinear Equations

Now consider the general *nonlinear* scalar ODE given by

$$\frac{du}{dt} = F(u, t) . \tag{6.83}$$

Application of the implicit Euler method gives

$$u_{n+1} = u_n + hF(u_{n+1}, t_{n+1}) . \tag{6.84}$$

This is a nonlinear difference equation. As an example, consider the nonlinear ODE

$$\frac{du}{dt} + \frac{1}{2}u^2 = 0 \tag{6.85}$$

solved using implicit Euler time marching, giving

$$u_{n+1} + h\frac{1}{2}u_{n+1}^2 = u_n , \tag{6.86}$$

which requires a nontrivial method to solve for u_{n+1}. There are several different approaches one can take to solving this nonlinear difference equation. An
iterative method, such as Newton's method (see below), can be used. In practice, the "initial guess" for this nonlinear problem can be quite close to the
solution, since the "initial guess" is simply the solution at the previous time
step, which implies that a linearization approach may be quite successful.
Such an approach is described in the next section.

6.10.3 Local Linearization for Scalar Equations

General Development. Let us start the process of local linearization by
considering (6.83). In order to implement the linearization, we expand $F(u,t)$
about some reference point in time. Designate the reference value by t_n and
the corresponding value of the dependent variable by u_n. A Taylor series
expansion about these reference quantities gives

$$F(u,t) = F(u_n, t_n) + \left(\frac{\partial F}{\partial u}\right)_n (u - u_n) + \left(\frac{\partial F}{\partial t}\right)_n (t - t_n)$$

$$+ \frac{1}{2}\left(\frac{\partial^2 F}{\partial u^2}\right)_n (u - u_n)^2 + \left(\frac{\partial^2 F}{\partial u \partial t}\right)_n (u - u_n)(t - t_n)$$

$$+ \frac{1}{2}\left(\frac{\partial^2 F}{\partial t^2}\right)_n (t - t_n)^2 + \cdots . \tag{6.87}$$

On the other hand, the expansion of $u(t)$ in terms of the independent variable
t is

$$u(t) = u_n + (t - t_n)\left(\frac{\partial u}{\partial t}\right)_n + \frac{1}{2}(t - t_n)^2 \left(\frac{\partial^2 u}{\partial t^2}\right)_n + \cdots . \tag{6.88}$$

If t is within h of t_n, both $(t - t_n)^k$ and $(u - u_n)^k$ are $O(h^k)$, and (6.87) can
be written as

$$F(u,t) = F_n + \left(\frac{\partial F}{\partial u}\right)_n (u - u_n) + \left(\frac{\partial F}{\partial t}\right)_n (t - t_n) + O(h^2) . \tag{6.89}$$

Notice that this is an expansion of the *derivative* of the function. Thus, relative to the order of expansion of the function, it represents a second-order-
accurate, locally-linear approximation to $F(u,t)$ that is valid in the vicinity
of the reference station t_n and the corresponding $u_n = u(t_n)$. With this we
obtain the locally (in the neighborhood of t_n) time-linear representation of
(6.83), namely

$$\frac{du}{dt} = \left(\frac{\partial F}{\partial u}\right)_n u + \left[F_n - \left(\frac{\partial F}{\partial u}\right)_n u_n\right] + \left(\frac{\partial F}{\partial t}\right)_n (t - t_n) + O(h^2) .$$

$$\tag{6.90}$$

Implementation of the Trapezoidal Method. As an example of how such an expansion can be used, consider the mechanics of applying the trapezoidal method for the time integration of (6.83). The trapezoidal method is given by

$$u_{n+1} = u_n + \frac{1}{2}h(F_{n+1} + F_n) + hO(h^2) , \tag{6.91}$$

where we write $hO(h^2)$ to emphasize that the method is second-order accurate. Using (6.89) to evaluate $F_{n+1} = F(u_{n+1}, t_{n+1})$, one finds

$$u_{n+1} = u_n + \frac{1}{2}h\left[F_n + \left(\frac{\partial F}{\partial u}\right)_n (u_{n+1} - u_n) + h\left(\frac{\partial F}{\partial t}\right)_n + O(h^2) + F_n\right]$$
$$+ hO(h^2) . \tag{6.92}$$

Note that the $O(h^2)$ term within the brackets (which is due to the local linearization) is multiplied by h and therefore is the same order as the $hO(h^2)$ error from the trapezoidal method. The use of local time linearization updated at the end of each time step and the trapezoidal time march combine to make a *second-order-accurate* numerical integration process. There are, of course, other second-order implicit time-marching methods that can be used. The important point to be made here is that *local linearization updated at each time step has not reduced the order of accuracy* of a second-order time-marching process.

A very useful reordering of the terms in (6.92) results in the expression

$$\left[1 - \frac{1}{2}h\left(\frac{\partial F}{\partial u}\right)_n\right]\Delta u_n = hF_n + \frac{1}{2}h^2\left(\frac{\partial F}{\partial t}\right)_n , \tag{6.93}$$

which is now in the delta form which will be formally introduced in Section 12.6. In many fluid mechanics applications the nonlinear function F is not an *explicit* function of t. In such cases the partial derivative of $F(u)$ with respect to t is zero, and (6.93) simplifies to the second-order accurate expression

$$\left[1 - \frac{1}{2}h\left(\frac{\partial F}{\partial u}\right)_n\right]\Delta u_n = hF_n . \tag{6.94}$$

Notice that the RHS is extremely simple. It is the product of h and the RHS of the basic equation evaluated at the previous time step. In this example, the basic equation was the simple scalar equation (6.83), but for our applications, it is generally the space-differenced form of the steady-state equation of some fluid flow problem.

A numerical time-marching procedure using (6.94) is usually implemented as follows:

1. Solve for the elements of $h\vec{F}_n$, store them in an array say $\vec{R}$, and save $\vec{u}_n$.

2. Solve for the elements of the matrix multiplying $\Delta \vec{u}_n$ and store in some appropriate manner making use of sparseness or bandedness of the matrix if possible. Let this storage area be referred to as B.
3. Solve the coupled set of linear equations

$$B \Delta \vec{u}_n = \vec{R}$$

for $\Delta \vec{u}_n$. (Very seldom does one find B^{-1} in carrying out this step.)
4. Find $\vec{u}_{n+1}$ by adding $\Delta \vec{u}_n$ to $\vec{u}_n$, thus

$$\vec{u}_{n+1} = \Delta \vec{u}_n + \vec{u}_n .$$

The solution for $\vec{u}_{n+1}$ is generally stored such that it overwrites the value of $\vec{u}_n$ and the process is repeated.

Implementation of the Implicit Euler Method. We have seen that the first-order implicit Euler method can be written as

$$u_{n+1} = u_n + h F_{n+1} . \tag{6.95}$$

If we introduce (6.90) into this method, rearrange terms, and remove the explicit dependence on time, we arrive at the form

$$\left[1 - h \left(\frac{\partial F}{\partial u} \right)_n \right] \Delta u_n = h F_n . \tag{6.96}$$

We see that the only difference between the implementation of the trapezoidal method and the implicit Euler method is the factor of $1/2$ in the brackets of the left side of (6.94) and (6.96). Omission of this factor degrades the method in time accuracy by one order of h. We shall see later that this method is an excellent choice for steady problems.

Newton's Method. Consider the limit $h \to \infty$ of (6.96) obtained by dividing both sides by h and setting $1/h = 0$. There results

$$-\left(\frac{\partial F}{\partial u} \right)_n \Delta u_n = F_n \tag{6.97}$$

or

$$u_{n+1} = u_n - \left[\left(\frac{\partial F}{\partial u} \right)_n \right]^{-1} F_n . \tag{6.98}$$

This is the well-known Newton method for finding the roots of the nonlinear equation $F(u) = 0$. The fact that it has quadratic convergence is verified by a glance at (6.87) and (6.88) (remember the dependence on t has been eliminated for this case). By quadratic convergence, we mean that the error after a given iteration is proportional to the square of the error at the previous iteration, where the error is the difference between the current solution and the converged solution. Quadratic convergence is thus a very powerful property. Use of a finite value of h in (6.96) leads to linear convergence, i.e. the error at a given iteration is some multiple of the error at the previous iteration. The reader should ponder the meaning of letting $h \to \infty$ for the trapezoidal method, given by (6.94).

6.10.4 Local Linearization for Coupled Sets of Nonlinear Equations

In order to present this concept, let us consider an example involving some simple boundary-layer equations. We choose the Falkner–Skan equations from classical boundary-layer theory. Our task is to apply the implicit trapezoidal method to the equations

$$\frac{d^3 f}{dt^3} + f\frac{d^2 f}{dt^2} + \beta\left[1 - \left(\frac{df}{dt}\right)^2\right] = 0 . \tag{6.99}$$

Here f represents a dimensionless stream function, and β is a scaling factor.

First of all we reduce (6.99) to a set of first-order nonlinear equations by the transformations

$$u_1 = \frac{d^2 f}{dt^2} \quad , \quad u_2 = \frac{df}{dt} \quad , \quad u_3 = f . \tag{6.100}$$

This gives the coupled set of three nonlinear equations

$$\begin{aligned} u_1' &= F_1 = -u_1 u_3 - \beta\left(1 - u_2^2\right) \\ u_2' &= F_2 = u_1 \\ u_3' &= F_3 = u_2 \end{aligned} \tag{6.101}$$

and these can be represented in vector notation as

$$\frac{d\vec{u}}{dt} = \vec{F}(\vec{u}) . \tag{6.102}$$

Now we seek to make the same local expansion that derived (6.90), except that this time we are faced with a nonlinear vector function, rather than a simple nonlinear scalar function. The required extension requires the evaluation of a matrix, called the Jacobian matrix.[9] Let us refer to this matrix as A. It is derived from (6.102) by the following process

$$A = (a_{ij}) = \partial F_i / \partial u_j . \tag{6.103}$$

For the general case involving a 3 × 3 matrix this is

---

[9] Recall that we derived the Jacobian matrices for the two-dimensional Euler equations in Section 2.2

$$A = \begin{bmatrix} \dfrac{\partial F_1}{\partial u_1} & \dfrac{\partial F_1}{\partial u_2} & \dfrac{\partial F_1}{\partial u_3} \\[2mm] \dfrac{\partial F_2}{\partial u_1} & \dfrac{\partial F_2}{\partial u_2} & \dfrac{\partial F_2}{\partial u_3} \\[2mm] \dfrac{\partial F_3}{\partial u_1} & \dfrac{\partial F_3}{\partial u_2} & \dfrac{\partial F_3}{\partial u_3} \end{bmatrix} . \tag{6.104}$$

The expansion of $\vec{F}(\vec{u})$ about some reference state $\vec{u}_n$ can be expressed in a way similar to the scalar expansion given by (6.87). Omitting the explicit dependency on the independent variable t, and defining $\vec{F}_n$ as $\vec{F}(\vec{u}_n)$, one has[10]

$$\vec{F}(\vec{u}) = \vec{F}_n + A_n(\vec{u} - \vec{u}_n) + O(h^2) , \tag{6.105}$$

where the argument for $O(h^2)$ is the same as in the derivation of (6.89). Using this we can write the local linearization of (6.102) as

$$\frac{d\vec{u}}{dt} = A_n\vec{u} + \underbrace{\left[\vec{F}_n - A_n\vec{u}_n \right]}_{\text{"constant"}} + O(h^2) , \tag{6.106}$$

which is a locally-linear, second-order-accurate approximation to a set of coupled nonlinear ordinary differential equations that is valid for $t \leq t_n + h$. Any first- or second-order time-marching method, explicit or implicit, could be used to integrate the equations without loss in accuracy with respect to order. The number of times, and the manner in which, the terms in the Jacobian matrix are updated as the solution proceeds depends, of course, on the nature of the problem.

Returning to our simple boundary-layer example, which is given by (6.101), we find the Jacobian matrix to be

$$A = \begin{bmatrix} -u_3 & 2\beta u_2 & -u_1 \\ 1 & 0 & 0 \\ 0 & 1 & 0 \end{bmatrix} , \tag{6.107}$$

The student should be able to derive results for this example that are equivalent to those given for the scalar case in (6.93). Thus for the Falkner–Skan equations the trapezoidal method results in

$$\begin{bmatrix} 1 + \frac{h}{2}(u_3)_n & -\beta h(u_2)_n & \frac{h}{2}(u_1)_n \\ -\frac{h}{2} & 1 & 0 \\ 0 & -\frac{h}{2} & 1 \end{bmatrix} \begin{bmatrix} (\Delta u_1)_n \\ (\Delta u_2)_n \\ (\Delta u_3)_n \end{bmatrix} = h \begin{bmatrix} -(u_1 u_3)_n - \beta(1 - u_2^2)_n \\ (u_1)_n \\ (u_2)_n \end{bmatrix} .$$

---

[10] The Taylor series expansion of a vector contains a vector for the first term, a matrix times a vector for the second term, and tensor products for the terms of higher order.

We find $\vec{u}_{n+1}$ from $\Delta\vec{u}_n + \vec{u}_n$, and the solution is now advanced one step. Re-evaluate the elements using $\vec{u}_{n+1}$ and continue. Without any iterating within a step advance, the solution will be second-order accurate in time.

Exercises

6.1 Find an expression for the nth term in the Fibonacci series, which is given by $1, 1, 2, 3, 5, 8, \ldots$ Note that the series can be expressed as the solution to a difference equation of the form $u_{n+1} = u_n + u_{n-1}$. What is u_{25}? (Let the first term given above be u_0.)

6.2 The trapezoidal method $u_{n+1} = u_n + \frac{1}{2}h(u'_{n+1} + u'_n)$ is used to solve the representative ODE.

(a) What is the resulting OΔE?
(b) What is its exact solution?
(c) How does the exact steady-state solution of the OΔE compare with the exact steady-state solution of the ODE if $\mu = 0$?

6.3 The 2nd-order backward method is given by

$$u_{n+1} = \frac{1}{3}\left(4u_n - u_{n-1} + 2hu'_{n+1}\right) .$$

(a) Write the OΔE for the representative equation. Identify the polynomials $P(E)$ and $Q(E)$.
(b) Derive the λ–σ relation. Solve for the σ-roots and identify them as principal or spurious.
(c) Find er_λ and the first two nonvanishing terms in a Taylor series expansion of the spurious root.

6.4 Consider the time-marching scheme given by

$$u_{n+1} = u_{n-1} + \frac{2h}{3}\left(u'_{n+1} + u'_n + u'_{n-1}\right) .$$

(a) Write the OΔE for the representative equation. Identify the polynomials $P(E)$ and $Q(E)$.
(b) Derive the λ–σ relation.
(c) Find er_λ.

6.5 Find the difference equation which results from applying the Gazdag predictor–corrector method (6.65) to the representative equation. Find the λ–σ relation.

6.6 Consider the following time-marching method:

$$\tilde{u}_{n+1/3} = u_n + hu'_n/3$$
$$\bar{u}_{n+1/2} = u_n + h\tilde{u}'_{n+1/3}/2$$
$$u_{n+1} = u_n + h\bar{u}'_{n+1/2}$$

Find the difference equation which results from applying this method to the representative equation. Find the λ–σ relation. Find the solution to the difference equation, including the homogeneous and particular solutions. Find er_λ and er_μ. What order is the homogeneous solution? What order is the particular solution? Find the particular solution if the forcing term is fixed.

6.7 Write a computer program to solve the one-dimensional linear convection equation with periodic boundary conditions and $a = 1$ on the domain $0 \le x \le 1$. Use 2nd-order centered differences in space and a grid of 50 points. For the initial condition, use

$$u(x,0) = e^{-0.5[(x-0.5)/\sigma]^2}$$

with $\sigma = 0.08$. Use the explicit Euler, 2nd-order Adams–Bashforth (AB2), implicit Euler, trapezoidal, and 4th-order Runge–Kutta methods. For the explicit Euler and AB2 methods, use a Courant number, $ah/\Delta x$, of 0.1; for the other methods, use a Courant number of unity. Plot the solutions obtained at $t = 1$ compared to the exact solution (which is identical to the initial condition).

6.8 Repeat Exercise 6.7 using 4th-order (noncompact) differences in space. Use only 4th-order Runge–Kutta time-marching at a Courant number of unity. Show solutions at $t = 1$ and $t = 10$ compared to the exact solution.

6.9 Using the computer program written for Exercise 6.7, compute the solution at $t = 1$ using 2nd-order centered differences in space coupled with the 4th-order Runge–Kutta method for grids of 100, 200, and 400 nodes. On a log–log scale, plot the error given by

$$\sqrt{\sum_{j=1}^{M} \frac{(u_j - u_j^{\text{exact}})^2}{M}} ,$$

where M is the number of grid nodes and u^{exact} is the exact solution. Find the global order of accuracy from the plot.

6.10 Using the computer program written for Exercise 6.8, repeat Exercise 6.9 using 4th-order (noncompact) differences in space.

6.11 Write a computer program to solve the one-dimensional linear convection equation with inflow–outflow boundary conditions and $a = 1$ on the domain $0 \leq x \leq 1$. Let $u(0,t) = \sin \omega t$ with $\omega = 10\pi$. Run until a periodic steady state is reached which is independent of the initial condition and plot your solution compared with the exact solution. Use 2nd-order centered differences in space with a 1st-order backward difference at the outflow boundary (as in (3.69)) together with 4th-order Runge–Kutta time marching. Use grids with 100, 200, and 400 nodes and plot the error vs. the number of grid nodes, as described in Exercise 6.9. Find the global order of accuracy.

6.12 Repeat Exercise 6.11 using 4th-order (noncompact) centered differences. Use a third-order forward-biased operator at the inflow boundary (as in (3.67)). At the last grid node, derive and use a 3rd-order backward operator (using nodes $j - 3$, $j - 2$, $j - 1$, and j) and at the second last node, use a 3rd-order backward-biased operator (using nodes $j - 2$, $j - 1$, j, and $j + 1$; see Exercise 3.1 in Chapter 3).

6.13 Using the approach described in Section 6.6.2, find the phase speed error, er_p, and the amplitude error, er_a, for the combination of second-order centered differences and 1st, 2nd, 3rd, and 4th-order Runge–Kutta time-marching at a Courant number of unity applied to the linear convection equation. Also plot the phase speed error obtained using exact integration in time, i.e. that obtained using the spatial discretization alone. Note that the required σ-roots for the various Runge–Kutta methods can be deduced from (6.69), without actually deriving the methods. Explain your results.

7. Stability of Linear Systems

A general definition of stability is neither simple nor universal and depends on the particular phenomenon being considered. In the case of the nonlinear ODE's of interest in fluid dynamics, stability is often discussed in terms of fixed points and attractors. In these terms a system is said to be stable in a certain domain if, from within that domain, some norm of its solution is always attracted to the same fixed point. These are important and interesting concepts, but we do not dwell on them in this work. Our basic concern is with time-dependent ODE's and OΔE's in which the coefficient matrices are independent of both u and t; see Section 4.2. We will refer to such matrices as *stationary*. Chapters 4 and 6 developed the representative forms of ODE's generated from the basic PDE's by the semi-discrete approach, and then the OΔE's generated from the representative ODE's by application of time-marching methods. These equations are represented by

$$\frac{d\vec{u}}{dt} = A\vec{u} - \vec{f}(t) \tag{7.1}$$

and

$$\vec{u}_{n+1} - C\vec{u}_n - \vec{g}_n . \tag{7.2}$$

respectively. For a one-step method, the latter form is obtained by applying a time-marching method to the generic ODE form in a fairly straightforward manner. For example, the explicit Euler method leads to $C = I + hA$, and $\vec{g}_n = h\vec{f}(nh)$. Methods involving two or more steps can always be written in the form of (7.2) by introducing new dependent variables. Note also that only methods in which the time and space discretizations are treated separately can be written in an intermediate semi-discrete form such as (7.1). The fully-discrete form, (7.2), and the associated stability definitions and analysis are applicable to all methods.

7.1 Dependence on the Eigensystem

Our definitions of stability are based entirely on the behavior of the homogeneous parts of (7.1) and (7.2). The stability of (7.1) depends entirely on the eigensystem[1] of A. The stability of (7.2) can often also be related to the eigensystem of its matrix. However, in this case the situation is not quite so simple since, in our applications to partial differential equations (especially hyperbolic ones), a stability definition can depend on both the time and space differencing. This is discussed in Section 7.4. Analysis of these eigensystems has the important added advantage that it gives an estimate of the *rate* at which a solution approaches a steady-state if a system is stable. Consideration will be given to matrices that have both complete and defective eigensystems, see Section 4.2.3, with a reminder that a complete system can be arbitrarily close to a defective one, in which case practical applications can make the properties of the latter appear to dominate.

If A and C are stationary, we can, in theory at least, estimate their fundamental properties. For example, in Section 4.3.2, we found from our model ODE's for diffusion and periodic convection what could be expected for the eigenvalue spectrums of practical physical problems containing these phenomena. These expectations are referred to many times in the following analysis of stability properties. They are important enough to be summarized by the following:

- For diffusion-dominated flows, the λ-eigenvalues tend to lie along the negative real axis.
- For periodic convection-dominated flows, the λ-eigenvalues tend to lie along the imaginary axis.

In many interesting cases, the eigenvalues of the matrices in (7.1) and (7.2) are sufficient to determine the stability. In previous chapters, we designated these eigenvalues as λ_m and σ_m for (7.1) and (7.2), respectively, and we will find it convenient to examine the stability of various methods in both the complex λ and complex σ planes.

7.2 Inherent Stability of ODE's

7.2.1 The Criterion

Here we state the standard stability criterion used for ordinary differential equations:

$$\boxed{\text{For a } \textit{stationary} \text{ matrix } A, \text{ (7.1) is } \textit{inherently stable} \text{ if, when } \vec{f} \text{ is constant, } \vec{u} \text{ remains bounded as } t \to \infty.} \qquad (7.3)$$

Note that inherent stability depends only on the transient solution of the ODE's.

---

[1] This is *not* the case if the coefficient matrix depends on t, even if it is linear.

7.2.2 Complete Eigensystems

If a matrix has a complete eigensystem, all of its eigenvectors are linearly independent, and the matrix can be diagonalized by a similarity transformation. In such a case it follows at once from (6.27), for example, that the ODE's are inherently stable if and only if

$$\boxed{\Re(\lambda_m) \leq 0 \quad \text{for all } m \,.}\tag{7.4}$$

This states that, for inherent stability, all of the λ eigenvalues must lie on, or to the left of, the imaginary axis in the complex λ plane. This criterion is satisfied for the model ODE's representing both diffusion and biconvection. It should be emphasized (as it is an important practical consideration in convection-dominated systems) that the special case for which $\lambda = \pm i$ is *included* in the domain of stability. In this case it is true that $\vec{u}$ does not decay as $t \to \infty$, but neither does it grow, so the above condition is met. Finally we note that for ODE's with complete eigensystems the eigenvectors play no role in the inherent stability criterion.

7.2.3 Defective Eigensystems

In order to understand the stability of ODE's that have defective eigensystems, we inspect the nature of their solutions in eigenspace. For this we draw on the results in Sections 4.2.3 and especially on (4.18) and (4.19) in that section. In an eigenspace related to defective systems the form of the representative equation changes from a single equation to a Jordan block. For example, instead of (4.45) a typical form of the homogeneous part might be

$$\begin{bmatrix} u_1' \\ u_2' \\ u_3' \end{bmatrix} = \begin{bmatrix} \lambda & & \\ 1 & \lambda & \\ & 1 & \lambda \end{bmatrix} \begin{bmatrix} u_1 \\ u_2 \\ u_3 \end{bmatrix},$$

for which one finds the solution

$$u_1(t) = u_1(0)e^{\lambda t}$$
$$u_2(t) = [u_2(0) + u_1(0)t]e^{\lambda t}$$
$$u_3(t) = \left[u_3(0) + u_2(0)t + \frac{1}{2}u_1(0)t^2 \right]e^{\lambda t} \,.\tag{7.5}$$

Inspecting this solution, we see that for such cases condition (7.4) must be modified to the form

$$\boxed{\Re(\lambda_m) < 0 \quad \text{for all } m}\tag{7.6}$$

since for pure imaginary λ, u_2 and u_3 would grow without bound (linearly or quadratically) if $u_2(0) \neq 0$ or $u_1(0) \neq 0$. Theoretically this condition is sufficient for stability in the sense of (7.3) since $t^k e^{-|\epsilon|t} \to 0$ as $t \to \infty$ for all non-zero ϵ. However, in practical applications the criterion may be worthless since there may be a very large growth of the polynomial before the exponential "takes over" and brings about the decay. Furthermore, on a computer such a growth might destroy the solution process before it could be terminated.

Note that the stability condition (7.6) *excludes* the imaginary axis which tends to be occupied by the eigenvalues related to biconvection problems. However, (7.6) is of little or no practical importance if significant amounts of dissipation are present.

7.3 Numerical Stability of OΔE's

7.3.1 The Criterion

The OΔE companion to (7.3) is:

> For a *stationary* matrix C, (7.2) is *numerically stable* if, when $\vec{g}$ is constant, $\vec{u}_n$ remains bounded as $n \to \infty$.
$$(7.7)$$

We see that numerical stability depends only on the transient solution of the OΔE's. This definition of stability is sometimes referred to as asymptotic or time stability.

As we stated at the beginning of this chapter, stability definitions are not unique. A definition often used in CFD literature stems from the development of PDE solutions that do not necessarily follow the semi-discrete route. In such cases it is appropriate to consider simultaneously the effects of both the time and space approximations. A time-space domain is fixed and stability is defined in terms of what happens to some norm of the solution within this domain as the mesh intervals go to zero at some constant ratio. We discuss this point of view in Section 7.4.

7.3.2 Complete Eigensystems

Consider a set of OΔE's governed by a complete eigensystem. The stability criterion, according to the condition set in (7.7), follows at once from a study of (6.28) and its companion for multiple σ-roots, (6.36). Clearly, for such systems a time-marching method is numerically stable if and only if

$$|(\sigma_m)_k| \le 1 \quad \text{for all } m \text{ and } k \,. \tag{7.8}$$

This condition states that, for numerical stability, all of the σ eigenvalues (both principal and spurious, if there are any) must lie on or inside the unit circle in the complex σ-plane.

This definition of stability for OΔE's is consistent with the stability definition for ODE's. Again the sensitive case occurs for the periodic-convection model which places the "correct" location of the principal σ-root precisely on the unit circle where the solution is only neutrally stable. Further, for a complete eigensystem, the eigenvectors play no role in the numerical stability assessment.

7.3.3 Defective Eigensystems

The discussion for these systems parallels the discussion for defective ODE's. Examine (6.18) and note its similarity with (7.5). We see that for defective OΔE's the required modification to (7.8) is

$$\boxed{|(\sigma_m)_k| < 1 \quad \text{for all } m \text{ and } k} \tag{7.9}$$

since defective systems do not guarantee boundedness for $|\sigma| = 1$. For example in (7.5) if $|\sigma| = 1$ and either $u_2(0)$ or $u_1(0) \neq 0$, we get linear or quadratic growth.

7.4 Time–Space Stability and Convergence of OΔE's

Let us now examine the concept of stability in a different way. In the previous discussion we considered in some detail the following approach:

(1) The PDE's are converted to ODE's by approximating the space derivatives on a finite mesh.
(2) Inherent stability of the ODE's is established by guaranteeing that $\Re(\lambda) \leq 0$.
(3) Time-march methods are developed which guarantee that $|\sigma(\lambda h)| \leq 1$ and this is taken to be the condition for numerical stability.

This *does* guarantee that a stationary system, generated from a PDE on some *fixed* space mesh, will have a numerical solution that is bounded as $t = nh \to \infty$. This *does not* guarantee that desirable solutions are generated in the time march process as both the time and space mesh intervals approach zero.

Now let us define stability in the time–space sense. First construct a finite time–space domain lying within $0 \leq x \leq L$ and $0 \leq t \leq T$. Cover this domain with a grid that is equispaced in both time and space and fix the mesh *ratio* by the equation[2]

$$c_n = \frac{\Delta t}{\Delta x} .$$

---

[2] This ratio is appropriate for hyperbolic problems; a different ratio may be needed for parabolic problems such as the model diffusion equation.

Next reduce our OΔE approximation of the PDE to a two-level (i.e. two time-planes) formula in the form of (7.2). The homogeneous part of this formula is

$$\vec{u}_{n+1} = C\vec{u}_n . \tag{7.10}$$

Equation (7.10) is said to be stable if any bounded initial vector, $\vec{u}_0$, produces a bounded solution vector, $\vec{u}_n$, as the mesh shrinks to zero for a fixed c_n. This is the classical definition of stability. It is often referred to as Lax or Lax–Richtmyer stability. Clearly as the mesh intervals go to zero, the number of time steps, N, must go to infinity in order to cover the entire fixed domain, so the criterion in (7.7) is a necessary condition for this stability criterion.

The significance of this definition of stability arises through *Lax's Theorem*, which states that, if a numerical method is *stable* (in the sense of Lax) and *consistent* then it is *convergent*. A method is *consistent* if it produces no error (in the Taylor series sense) in the limit as the mesh spacing and the time step go to zero (with c_n fixed, in the hyperbolic case). This is further discussed in Section 7.8. A method is *convergent* if it converges to the exact solution as the mesh spacing and time step go to zero in this manner.[3] Clearly, this is an important property.

Applying simple recursion to (7.10), we find

$$\vec{u}_n = C^n \vec{u}_0$$

and using vector and matrix p-norms (see Appendix A) and their inequality relations, we have

$$||\vec{u}_n|| = ||C^n \vec{u}_0|| \leq ||C^n|| \cdot ||\vec{u}_0|| \leq ||C||^n \cdot ||\vec{u}_0|| . \tag{7.11}$$

Since the initial data vector is bounded, the solution vector is bounded if

$$||C|| \leq 1 , \tag{7.12}$$

where $||C||$ represents any p-norm of C. This is often used as a *sufficient* condition for stability.

Now we need to relate the stability definitions given in (7.8) and (7.9) with that given in (7.12). In (7.8) and (7.9), stability is related to the *spectral radius* of C, i.e. its eigenvalue of maximum magnitude. In (7.12), stability is related to a p-norm of C. It is clear that *the criteria are the same when the spectral radius is a true p-norm.*

Two facts about the relation between spectral radii and matrix norms are well known:

---

[3] In the CFD literature, the word *converge* is used with two entirely different meanings. Here we refer to a numerical solution *converging* to the exact solution of the PDE. Later we will refer to the numerical solution *converging* to a steady-state solution.

(1) The spectral radius of a matrix is its L_2 norm when the matrix is normal, i.e. it commutes with its transpose.
(2) The spectral radius is the *lower bound* of all norms.

Furthermore, when C is normal, the second inequality in (7.11) becomes an equality. In this case, (7.12) becomes both necessary and sufficient for stability. From these relations we draw two important conclusions about the numerical stability of methods used to solve PDE's.

- The stability criteria in (7.8) and (7.12) are identical for stationary systems when the governing matrix is normal. This includes symmetric, asymmetric, and circulant matrices. These criteria are both necessary and sufficient for methods that generate such matrices and depend solely upon the eigenvalues of the matrices.
- If the spectral radius of *any* governing matrix is greater than one, the method is unstable by any criterion. Thus for general matrices, the spectral radius condition is necessary[4] but not sufficient for stability.

7.5 Numerical Stability Concepts in the Complex σ-Plane

7.5.1 σ-Root Traces Relative to the Unit Circle

Whether or not the semi-discrete approach was taken to find the differencing approximation of a set of PDE's, the final difference equations can be represented by

$$\vec{u}_{n+1} = C\vec{u}_n - \vec{g}_n \ .$$

Furthermore if C has a complete[5] eigensystem, the solution to the homogeneous part can always be expressed as

$$\vec{u}_n = c_1 \sigma_1^n \vec{x}_1 + \cdots + c_m \sigma_m^n \vec{x}_m + \cdots + c_M \sigma_M^n \vec{x}_M \ ,$$

where the σ_m are the eigenvalues of C. If the semi-discrete approach *is* used, we can find a relation between the σ and the λ eigenvalues. This serves as a very convenient guide as to where we might expect the σ-roots to lie relative to the unit circle in the complex σ-plane. For this reason we will proceed to trace the locus of the σ-roots as a function of the parameter λh for the equations modeling diffusion and periodic convection[6].

---

[4] Actually the necessary condition is that the spectral radius of C be less than or equal to $1 + O(\Delta t)$, but this distinction is not critical for our purposes here.
[5] The subject of defective eigensystems has been addressed. From now on we will omit further discussion of this special case.
[6] Or, if you like, the parameter h for fixed values of λ equal to -1 and i for the diffusion and biconvection cases, respectively.

Locus of the Exact Trace. Figure 7.1 shows the exact trace of the σ-root if it is generated by $e^{\lambda h}$ representing either diffusion or biconvection. In both cases the • represents the starting value where $h = 0$ and $\sigma = 1$. For the diffusion model, λh is real and negative. As the magnitude of λh increases, the trace representing the dissipation model heads towards the origin as $\lambda h = \rightarrow -\infty$. On the other hand, for the biconvection model, $\lambda h = i\omega h$ is always imaginary. As the magnitude of ωh increases, the trace representing the biconvection model travels around the circumference of the unit circle, which it never leaves. We must be careful in interpreting σ when it is representing $e^{i\omega h}$. The fact that it *lies on* the unit circle means only that the *amplitude* of the representation is correct, it tells us nothing of the *phase* error (see (6.41)). The phase error relates to the *position on* the unit circle.

Examples of Some Methods. Now let us compare the exact σ-root traces with some that are produced by actual time-marching methods. Table 7.1 shows the λ–σ relations for a variety of methods. Figure 7.2 illustrates the results produced by various methods when they are applied to the model ODE's for diffusion and periodic convection, (4.4) and (4.5). It is implied that the behavior shown is typical of what will happen if the methods are applied to diffusion- (or dissipation-) dominated or periodic convection-dominated problems as well as what does happen in the model cases. Most of the important possibilities are covered by the illustrations.

Table 7.1. Some λ–σ relations

1.	$\sigma - 1 - \lambda h = 0$	Explicit Euler
2.	$\sigma^2 - 2\lambda h\sigma - 1 = 0$	Leapfrog
3.	$\sigma^2 - (1 + \frac{3}{2}\lambda h)\sigma + \frac{1}{2}\lambda h = 0$	AB2
4.	$\sigma^3 - (1 + \frac{23}{12}\lambda h)\sigma^2 + \frac{16}{12}\lambda h\sigma - \frac{5}{12}\lambda h = 0$	AB3
5.	$\sigma(1 - \lambda h) - 1 = 0$	Implicit Euler
6.	$\sigma(1 - \frac{1}{2}\lambda h) - (1 + \frac{1}{2}\lambda h) = 0$	Trapezoidal
7.	$\sigma^2(1 - \frac{2}{3}\lambda h) - \frac{4}{3}\sigma + \frac{1}{3} = 0$	2nd-order Backward
8.	$\sigma^2(1 - \frac{5}{12}\lambda h) - (1 + \frac{8}{12}\lambda h)\sigma + \frac{1}{12}\lambda h = 0$	AM3
9.	$\sigma^2 - (1 + \frac{13}{12}\lambda h + \frac{15}{24}\lambda^2 h^2)\sigma + \frac{1}{12}\lambda h(1 + \frac{5}{2}\lambda h) = 0$	ABM3
10.	$\sigma^3 - (1 + 2\lambda h)\sigma^2 + \frac{3}{2}\lambda h\sigma - \frac{1}{2}\lambda h = 0$	Gazdag
11.	$\sigma - 1 - \lambda h - \frac{1}{2}\lambda^2 h^2 = 0$	RK2
12.	$\sigma - 1 - \lambda h - \frac{1}{2}\lambda^2 h^2 - \frac{1}{6}\lambda^3 h^3 - \frac{1}{24}\lambda^4 h^4 = 0$	RK4
13.	$\sigma^2(1 - \frac{1}{3}\lambda h) - \frac{4}{3}\lambda h\sigma - (1 + \frac{1}{3}\lambda h) = 0$	Milne 4th

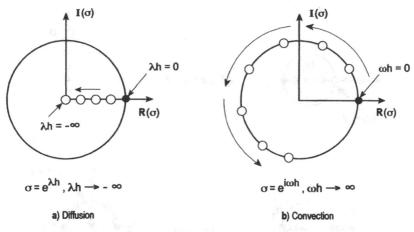

a) Diffusion   b) Convection

Fig. 7.1. Exact traces of σ-roots for model equations

(a) **Explicit Euler Method.** Figure 7.2a shows results for the explicit Euler method. When used for diffusion-dominated cases, it is stable for the range $-2 \leq \lambda h \leq 0$. (Usually the magnitude of λ has to be estimated, and often it is found by trial and error). When used for biconvection, the σ-trace falls outside the unit circle for all finite h, and the method has no range of stability in this case.

(b) **Leapfrog Method.** This is a two-root method, since there are two σ's produced by every λ. When applied to diffusion-dominated problems, we see from Figure 7.2b that the principal root is stable for a range of λh, but the spurious root is not. In fact, the spurious root starts on the unit circle and falls outside of it for *all* $\Re(\lambda h) < 0$. However, for biconvection cases, when λ is *pure imaginary*, the method is not only stable, but it also produces a σ that falls precisely on the unit circle in the range $0 \leq \omega h \leq 1$. As was pointed out above, this does not mean that the method is without error. Although the figure shows that there is a range of ωh in which the leapfrog method produces no error in amplitude, it says nothing about the error in *phase*. More is said about this in Chapter 8.

(c) **Second-Order Adams–Bashforth Method.** This is also a two-root method but, unlike the leapfrog scheme, the spurious root starts at the origin, rather than on the unit circle (see Figure 7.2c). Therefore, there is a range of real negative λh for which the method will be stable. The figure shows that the range ends when $\lambda h < -1.0$, since at that point the spurious root leaves the circle, and $|\sigma_2|$ becomes greater than one. The situation is quite different when the λ-root is pure imaginary. In that case as ωh increases away from zero, the spurious root remains inside the circle and remains stable for a range of ωh. However, the principal root falls outside the unit circle for all

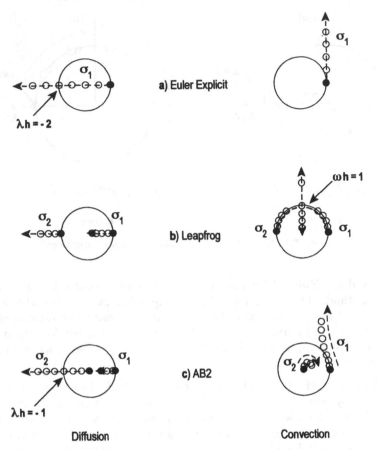

a) Euler Explicit

b) Leapfrog

c) AB2

$\lambda h = -2$

$\lambda h = -1$

$\omega h = 1$

σ_1

σ_2

Diffusion   Convection

Fig. 7.2. a–c Traces of σ-roots for various methods

$\omega h > 0$, and for the biconvection model equation the method is unstable for all h.

(d) Trapezoidal Method. The trapezoidal method is a very popular one for reasons that are partially illustrated in Figure 7.2d. Its σ-roots fall on or inside the unit circle for both the diffusing and the periodic convecting case and, in fact, *it is stable for all values of λh for which λ itself is inherently stable*. Just like the leapfrog method, it has the capability of producing only phase error for the periodic convecting case, but there is a major difference between the two since the trapezoidal method produces no amplitude error for *any* ωh – not just a limited range between $0 \leq \omega h \leq 1$.

(e) Gazdag Method. The Gazdag method was designed to produce low phase error. Since its characteristic polynomial for σ is a cubic (Table 7.1, no. 10), it must have two spurious roots in addition to the principal one. These are shown in Figure 7.2e. In both the diffusion and biconvection cases,

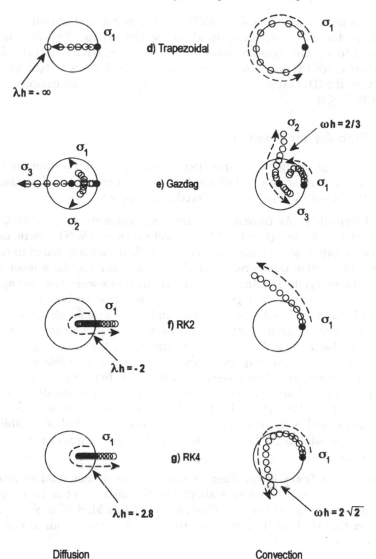

Fig. 7.2. d–g Traces of σ-roots for various methods

a spurious root limits the stability. For the diffusing case, a spurious root leaves the unit circle when $\lambda h < -1/2$, and for the biconvecting case, when $\omega h > 2/3$. Note that both spurious roots are located at the origin when $\lambda = 0$.

(f, g) Second- and Fourth-Order Runge–Kutta Methods, RK2 and RK4. Traces of the σ-roots for the second- and fourth-order Runge–Kutta methods are shown in Figures 7.2f and g. The figures show that both methods are stable for a range of λh when λh is real and negative, but that the range of stability for RK4 is greater, going almost all the way to -2.8, whereas RK2 is limited to -2. On the other hand, for biconvection, RK2 is unstable for all ωh, whereas RK4 remains inside the unit circle for $0 \le \omega h \le 2\sqrt{2}$. One can show that the RK4 stability limit is about $|\lambda h| < 2.8$ for all complex λh for which $\Re(\lambda) \le 0$.

7.5.2 Stability for Small Δt

It is interesting to pursue the question of stability when the time step size, h, is small, so that accuracy of all the λ-roots is of importance. Situations for which this is not the case are considered in Chapter 8.

Mild Instability. All conventional time-marching methods produce a principal root that is very close to $e^{\lambda h}$ for small values of λh. Therefore, on the basis of the principal root, the stability of a method that is required to resolve a transient solution over a relatively short time span may be a moot issue. Such cases are typified by the AB2 and RK2 methods when they are applied to a biconvection problem. Figures 7.2c and f show that for both methods the principal root falls outside the unit circle and is unstable for all ωh. However, if the transient solution of interest can be resolved in a limited number of time steps that are small in the sense of the figure, the error caused by this instability may be relatively unimportant. If the root had fallen inside the circle the method would have been declared stable but an error of the same *magnitude* would have been committed, just in the opposite direction. For this reason the AB2 and the RK2 methods have both been used in serious quantitative studies involving periodic convection. This kind of instability is referred to as mild instability and is not a serious problem under the circumstances discussed.

Catastrophic Instability. There is a much more serious stability problem for small h that can be brought about by the existence of certain types of spurious roots. One of the best illustrations of this kind of problem stems from a critical study of the most accurate, explicit, two-step, linear multistep method (see Table 7.1)

$$u_{n+1} = -4u_n + 5u_{n-1} + 2h\left(2u_n' + u_{n-1}'\right) . \tag{7.13}$$

One can show, using the methods given in Section 6.6, that this method is third-order accurate both in terms of er_λ and er_μ, so from an accuracy point of view it is attractive. However, let us inspect its stability even for very small values of λh. This can easily be accomplished by studying its characteristic polynomial when $\lambda h \to 0$. From (7.13) it follows that for $\lambda h = 0$, $P(E) = E^2 + 4E - 5$. Factoring $P(\sigma) = 0$ we find $P(\sigma) = (\sigma - 1)(\sigma + 5) = 0$. There are two σ-roots: σ_1, the principal one, equal to 1, and σ_2, a spurious one, equal to -5!

In order to evaluate the consequences of this result, one must understand how methods with spurious roots work in practice. We know that they are not self-starting, and the special procedures chosen to start them initialize the coefficients of the spurious roots, the c_{mk} for $k > 1$ in (6.36). If the starting process is well designed, these coefficients are forced to be very small, and if the method is stable, they get smaller with increasing n. However, if the magnitude of one of the spurious σ is equal to 5, one can see disaster is imminent because $(5)^{10} \approx 10^7$. Even a very small initial value of c_{mk} is quickly overwhelmed. Such methods are called catastrophically unstable and are worthless for most, if not all, computations.

Milne and Adams Methods. If we inspect the σ-root traces of the multiple root methods in Figure 7.2, we find them to be of two types. One type is typified by the leapfrog method. In this case a spurious root falls *on the unit circle* when $h \rightarrow 0$. The other type is exemplified by the 2nd-order Adams–Bashforth and Gazdag methods. In this case all spurious roots fall *on the origin* when $h \rightarrow 0$.

The former type is referred to as a *Milne method*. Since at least one spurious root for a Milne method always starts on the unit circle, the method is likely to become unstable for some complex λ as h proceeds away from zero. On the basis of a Taylor series expansion, however, these methods are generally the most accurate insofar as they minimize *the coefficient* in the leading term for er_t.

The latter type is referred to as an *Adams method*. Since for these methods all spurious methods start at the origin for $h = 0$, they have a guaranteed range of stability for small enough h. However, on the basis of the magnitude of the coefficient in the leading Taylor series error term, they suffer, relatively speaking, from accuracy.

For a given amount of computational work, the order of accuracy of the two types is generally equivalent, and stability requirements in CFD applications generally override the (usually small) increase in accuracy provided by a coefficient with lower magnitude.

7.6 Numerical Stability Concepts in the Complex λh Plane

7.6.1 Stability for Large h

The reason to study stability for small values of h is fairly easy to comprehend. Presumably we are seeking to resolve some transient and, since the accuracy of the transient solution for all of our methods depends on the smallness of the time-step, we seek to make the size of this step as small as possible. On the other hand, the cost of the computation generally depends on the number of steps taken to compute a solution, and to minimize this we wish to make the step size as large as possible. In the compromise, stability can play a part. Aside from ruling out catastrophically unstable methods, however, the situation in which all of the transient terms are resolved constitutes a rather minor role in stability considerations.

By far the most important aspect of numerical stability occurs under conditions when:

- one has inherently stable, coupled systems with λ-eigenvalues having widely separated magnitudes,

or

- we seek only to find a steady-state solution using a path that includes the unwanted transient.

In both of these cases there exist in the eigensystems relatively large values of $|\lambda h|$ associated with eigenvectors that we wish to drive through the solution process without any regard for their individual accuracy in eigenspace. This situation is the major motivation for the study of numerical stability. It leads to the subject of stiffness discussed in the next chapter.

7.6.2 Unconditional Stability, *A*-Stable Methods

Inherent stability of a set of ODE's was defined in Section 7.2 and, for coupled sets with a complete eigensystem, it amounted to the requirement that the real parts of all λ eigenvalues must lie on, or to the left of, the imaginary axis in the complex λ plane. This serves as an excellent reference frame to discuss and define the general stability features of time-marching methods. For example, we start with the definition:

> A numerical method is *unconditionally stable* if it is stable for all ODE's that are inherently stable.

A method with this property is said to be *A-stable*. A method is A_0-*stable* if the region of stability contains the negative real axis in the complex λh plane, and *I-stable* if it contains the entire imaginary axis. By applying a fairly simple test for *A*-stability in terms of positive real functions to the class of two-step LMM's given in Section 6.7.3, one finds these methods to be *A*-stable if and only if

$$\theta \geq \varphi + \frac{1}{2} \tag{7.14}$$

$$\xi \geq -\frac{1}{2} \tag{7.15}$$

$$\xi \leq \theta + \varphi - \frac{1}{2} . \tag{7.16}$$

A set of *A*-stable implicit methods is shown in Table 7.2.

Table 7.2. Some unconditionally stable (A-stable) implicit methods

θ	ξ	φ	Method	Order
1	0	0	Implicit Euler	1
1/2	0	0	Trapezoidal	2
1	1/2	0	2nd O Backward	2
3/4	0	−1/4	Adams type	2
1/3	−1/2	−1/3	Lees	2
1/2	−1/2	−1/2	Two-step trapezoidal	2
5/8	−1/6	−2/9	A-contractive	2

Notice that none of these methods has an accuracy higher than second-order. It can be proved that the order of an A-stable LMM *cannot exceed two*, and, furthermore that of all 2nd-order A-stable methods, the trapezoidal method has the smallest truncation error.

Returning to the stability test using positive real functions, one can show that a two-step LMM is A_0-stable if and only if

$$\theta \geq \varphi + \frac{1}{2} \tag{7.17}$$

$$\xi \geq -\frac{1}{2} \tag{7.18}$$

$$0 \leq \theta - \varphi . \tag{7.19}$$

For first-order accuracy, the inequalities (7.17)–(7.19) are less stringent than (7.14)–(7.16). For second-order accuracy, however, the parameters (θ, ξ, φ) are related by the condition

$$\varphi - \xi - \theta + \frac{1}{2}$$

and the two sets of inequalities reduce to the same set which is

$$\xi \leq 2\theta - 1 \tag{7.20}$$

$$\xi \geq -\frac{1}{2} . \tag{7.21}$$

Hence, two-step, second-order accurate LMM's that are A-stable and A_0-stable share the same (φ, ξ, θ) parameter space. Although the order of accuracy of an A-stable method cannot exceed two, A_0-stable LMM methods exist which have an accuracy of arbitrarily high order.

It has been shown that for a method to be I-stable it must also be A-stable. Therefore, no further discussion is necessary for the special case of I-stability.

It is not difficult to prove that methods having a characteristic polynomial for which the coefficient of the highest order term in E is unity[7] can never be unconditionally stable. This includes all explicit methods and predictor–corrector methods made up of explicit sequences. Such methods are referred to, therefore, as *conditionally stable* methods.

7.6.3 Stability Contours in the Complex λh Plane

A very convenient way to present the stability properties of a time-marching method is to plot the locus of the complex λh for which $|\sigma| = 1$, such that the resulting contour goes through the point $\lambda h = 0$. Here $|\sigma|$ refers to the maximum absolute value of any σ, principal or spurious, that is a root to the characteristic polynomial for a given λh. It follows from Section 7.3 that on one side of this contour the numerical method is stable, while on the other, it is unstable. We refer to it, therefore, as a *stability contour*.

Typical stability contours for both explicit and implicit methods are illustrated in Figure 7.3, which is derived from the one-root θ-method given in Section 6.5.4.

Contours for Explicit Methods. Figure 7.3a shows the stability contour for the explicit Euler method. In the following two ways it is typical of all stability contours for explicit methods:

(1) The contour encloses a finite portion of the left-half complex λh-plane.
(2) The region of stability is *inside* the boundary, and therefore, it is conditional.

However, this method includes no part of the imaginary axis (except for the origin) and so it is unstable for the model biconvection problem. Although several explicit methods share this deficiency (e.g. AB2, RK2), several others do not (e.g. leapfrog, Gazdag, RK3, RK4) (see Figures 7.4 and 7.5). Notice in particular that the third- and fourth-order Runge–Kutta methods, Figure 7.5, include a portion of the imaginary axis out to $\pm 1.9i$ and $\pm 2\sqrt{2}i$, respectively.

Contours for Unconditionally Stable Implicit Methods. Figure 7.3c shows the stability contour for the implicit Euler method. It is typical of many stability contours for unconditionally stable implicit methods. Notice that the method is stable for the entire range of complex λh that fall *outside* the boundary. This means that the method is numerically stable *even when the ODE's that it is being used to integrate are inherently unstable.* Some other implicit unconditionally stable methods with the same property are shown in Figure 7.6. In all of these cases the imaginary axis is part of the *stable* region.

---

[7] Or can be made equal to unity by a trivial normalization (division by a constant independent of λh). The proof follows from the fact that the coefficients of such a polynomial are sums of various combinations of products of all its roots.

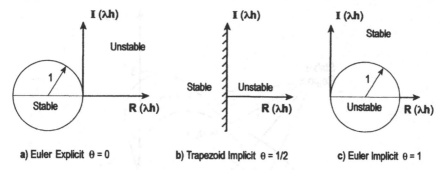

a) Euler Explicit $\theta = 0$ b) Trapezoid Implicit $\theta = 1/2$ c) Euler Implicit $\theta = 1$

Fig. 7.3. Stability contours for the θ-method

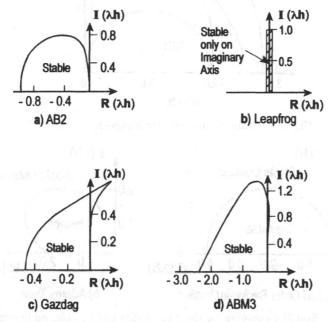

Fig. 7.4. Stability contours for some explicit methods

Not all unconditionally stable methods are stable in some regions where the ODE's they are integrating are inherently unstable. The classic example of a method that is stable *only* when the generating ODE's are themselves inherently stable is the trapezoidal method, i.e. the special case of the θ-method for which $\theta = 1/2$. The stability boundary for this case is shown in Figure 7.3b. The boundary is the imaginary axis and the numerical method is stable for λh lying on or to the left of this axis. Two other methods that have this property are the two-step trapezoidal method

$$u_{n+1} = u_{n-1} + h\left(u'_{n+1} + u'_{n-1}\right)$$

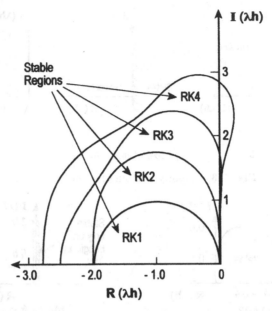

Fig. 7.5. Stability contours for Runge–Kutta methods

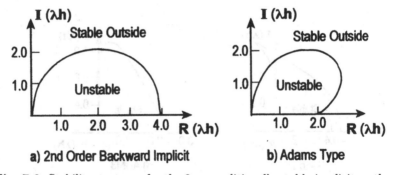

Fig. 7.6. Stability contours for the 2 unconditionally stable implicit methods

and a method due to Lees

$$u_{n+1} = u_{n-1} + \frac{2}{3}h\left(u'_{n+1} + u'_n + u'_{n-1}\right) .$$

Notice that both of these methods are of the Milne type.

Contours for Conditionally Stable Implicit Methods. Just because a method is implicit does not mean that it is unconditionally stable. Two illustrations of this are shown in Figure 7.7. One of these is the Adams–Moulton 3rd-order method (Table 7.1, no. 8). Another is the 4th-order Milne method given by the point operator

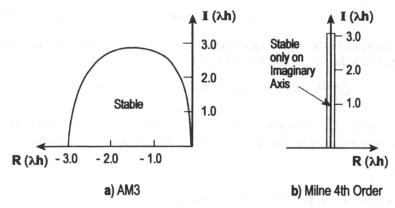

Fig. 7.7. Stability contours for two conditionally stable implicit methods

$$u_{n+1} = u_{n-1} + \frac{1}{3}h\left(u'_{n+1} + 4u'_n + u'_{n-1}\right)$$

and shown in Table 7.1, no. 13. It is stable *only* for $\lambda = \pm i\omega$ when $0 \leq \omega \leq \sqrt{3}$. Its stability boundary is very similar to that for the leapfrog method (see Figure 7.4b).

7.7 Fourier Stability Analysis

By far the most popular form of stability analysis for numerical schemes is the Fourier or von Neumann approach. This analysis is usually carried out on point operators and it does not depend on an intermediate stage of ODE's. *Strictly speaking* it applies only to difference approximations of PDE's that produce OΔE's *which are linear, have no space or time varying coefficients, and have periodic boundary conditions.*[8] In practical application it is often used as a guide for estimating the worthiness of a method for more general problems. It serves as a fairly reliable *necessary* stability condition, but it is by no means a *sufficient* one.

7.7.1 The Basic Procedure

One takes data from a "typical" point in the flow field and uses this as constant throughout time and space according to the assumptions given above. Then one imposes a spatial harmonic as an initial value on the mesh and asks the question: Will its amplitude grow or decay in time? The answer is determined by finding the conditions under which

---

[8] Another way of viewing this is to consider it as an initial value problem on an infinite space domain.

$$u(x,t) = e^{\alpha t} \cdot e^{i\kappa x} \tag{7.22}$$

is a solution to the *difference* equation, where κ is real and $\kappa \Delta x$ lies in the range $0 \le \kappa \Delta x \le \pi$. Since, for the general term,

$$u_{j+m}^{(n+\ell)} = e^{\alpha(t+\ell \Delta t)} \cdot e^{i\kappa(x+m\Delta x)} = e^{\alpha \ell \Delta t} \cdot e^{i\kappa m \Delta x} \cdot u_j^{(n)}$$

the quantity $u_j^{(n)}$ is common to every term and can be factored out. In the remaining expressions, we find the term $e^{\alpha \Delta t}$, which we represent by σ, thus

$$\sigma \equiv e^{\alpha \Delta t} .$$

Then, since $e^{\alpha t} = \left(e^{\alpha \Delta t}\right)^n = \sigma^n$, it is clear that:

$$\boxed{\text{For numerical stability } |\sigma| \le 1} \tag{7.23}$$

and the problem is to solve for the σ's produced by any given method and, as a necessary condition for stability, make sure that, in the worst possible combination of parameters, (7.23) is satisfied.[9]

7.7.2 Some Examples

The procedure can best be explained by examples. Consider as a first example the finite-difference approximation to the model diffusion equation known as Richardson's method of overlapping steps. This was mentioned in Chapter 4 and given as (4.2):

$$u_j^{(n+1)} = u_j^{(n-1)} + \nu \frac{2\Delta t}{\Delta x^2}\left(u_{j+1}^{(n)} - 2u_j^{(n)} + u_{j-1}^{(n)}\right) . \tag{7.24}$$

Substitution of (7.22) into (7.24) gives the relation

$$\sigma = \sigma^{-1} + \nu \frac{2\Delta t}{\Delta x^2}\left(e^{i\kappa \Delta x} - 2 + e^{-i\kappa \Delta x}\right)$$

or

$$\sigma^2 + \underbrace{\left[\frac{4\nu \Delta t}{\Delta x^2}(1 - \cos\kappa \Delta x)\right]}_{2b}\sigma - 1 = 0 . \tag{7.25}$$

Thus (7.22) is a solution of (7.24) if σ is a root of (7.25). The two roots of (7.25) are

$$\sigma_{1,2} = -b \pm \sqrt{b^2 + 1} ,$$

---

[9] If boundedness is required in a *finite* time domain, the condition is often presented as $|\sigma| \le 1 + O(\Delta t)$.

from which it is clear that one $|\sigma|$ is always > 1. We find, therefore, that by the Fourier stability test, Richardson's method of overlapping steps is unstable for all ν, κ and Δt.

As another example consider the finite-difference approximation for the model biconvection equation

$$u_j^{(n+1)} = u_j^{(n)} - \frac{a\Delta t}{2\Delta x}\left(u_{j+1}^{(n)} - u_{j-1}^{(n)}\right) . \tag{7.26}$$

In this case

$$\sigma = 1 - \frac{a\Delta t}{\Delta x} \cdot \mathrm{i} \cdot \sin \kappa \Delta x ,$$

from which it is clear that $|\sigma| > 1$ for all nonzero a and κ. Thus we have another finite-difference approximation that, by the Fourier stability test, is unstable for any choice of the free parameters.

7.7.3 Relation to Circulant Matrices

The underlying assumption in a Fourier stability analysis is that the C matrix, determined when the differencing scheme is put in the form of (7.2), is circulant. Such being the case, the term $\mathrm{e}^{\mathrm{i}\kappa x}$ in (7.22) represents an *eigenvector* of the system, and the two examples just presented outline a simple procedure for finding the *eigenvalues* of the circulant matrices formed by application of the two methods to the model problems. The choice of σ for the stability parameter in the Fourier analysis, therefore, is not an accident. It is exactly the same σ we have been using in all of our previous discussions, but arrived at from a different perspective.

If we examine the preceding examples from the viewpoint of circulant matrices and the semi-discrete approach, the results present rather obvious conclusions. The space differencing in Richardson's method produces the matrix $s \cdot B_\mathrm{p}(1, -2, 1)$ where s is a positive scalar coefficient. From Appendix B we find that the eigenvalues of this matrix are real negative numbers. Clearly, the time-marching is being carried out by the leapfrog method and, from Figure 7.4, this method is unstable for *all* eigenvalues with negative real parts. On the other hand, the space matrix in (7.26) is $B_\mathrm{p}(-1, 0, 1)$, and according to Appendix B, this matrix has pure imaginary eigenvalues. However, in this case the explicit Euler method is being used for the time-march and, according to Figure 7.3, this method is always unstable for such conditions.

7.8 Consistency

Consider the model equation for diffusion analysis

$$\frac{\partial u}{\partial t} = \nu \frac{\partial^2 u}{\partial x^2} . \tag{7.27}$$

Many years before computers became available (1910, in fact), Lewis F. Richardson proposed a method for integrating equations of this type. We presented his method in (4.2) and analyzed its stability by the Fourier method in Section 7.7.

In Richardson's time, the concept of numerical instability was not known. However, the concept is quite clear today and we now know immediately that his approach would be unstable. As a semi-discrete method it can be expressed in matrix notation as the system of ODE's:

$$\frac{d\vec{u}}{dt} = \frac{\nu}{\Delta x^2} B(1, -2, 1)\vec{u} + (\vec{bc}) \tag{7.28}$$

with the leapfrog method used for the time march. Our analysis in this chapter revealed that this is numerically unstable since the λ-roots of $B(1, -2, 1)$ are all real and negative and the spurious σ-root in the leapfrog method is unstable for all such cases (see Figure 7.4b).

The method was used by Richardson for weather prediction, and this fact can now be a source of some levity. In all probability, however, the hand calculations (the only approach available at the time) were not carried far enough to exhibit strange phenomena. We could, of course, use the 2nd-order Runge–Kutta method to integrate (7.28) since it is stable for real negative λ's. It is, however, conditionally stable and for this case we are rather severely limited in time step size by the requirement $\Delta t \leq \Delta x^2/(2\nu)$.

There are many ways to manipulate the numerical stability of algorithms. One of them is to introduce mixed time and space differencing, a possibility we have not yet considered. For example, we introduced the DuFort–Frankel method in Chapter 4:

$$u_j^{(n+1)} = u_j^{(n-1)} + \frac{2\nu\Delta t}{\Delta x^2}\left[u_{j-1}^{(n)} - 2\left(\frac{u_j^{(n+1)} + u_j^{(n-1)}}{2}\right) + u_{j+1}^{(n)}\right], \tag{7.29}$$

in which the central term in the space derivative in (4.2) has been replaced by its average value at two different time levels. Now let

$$\alpha \equiv \frac{2\nu\Delta t}{\Delta x^2}$$

and rearrange terms

$$(1+\alpha)u_j^{(n+1)} = (1-\alpha)u_j^{(n-1)} + \alpha\left(u_{j-1}^{(n)} + u_{j+1}^{(n)}\right).$$

There is no obvious ODE between the basic PDE and this final OΔE. Hence, there is no intermediate λ-root structure to inspect. Instead one proceeds immediately to the σ-roots.

The simplest way to carry this out is by means of the Fourier stability analysis introduced in Section 7.7. This leads at once to

$$(1 + \alpha)\sigma = (1 - \alpha)\sigma^{-1} + \alpha\left(e^{i\kappa\Delta x} + e^{-i\kappa\Delta x}\right)$$

or

$$(1 + \alpha)\sigma^2 - 2\alpha\sigma\cos(k\Delta x) - (1 - \alpha) = 0 .$$

The solution of the quadratic is

$$\sigma = \frac{\alpha\cos\kappa\Delta x \pm \sqrt{1 - \alpha^2\sin^2\kappa\Delta x}}{1 + \alpha} .$$

There are $2M$ σ-roots all of which are ≤ 1 for any real α in the range $0 \leq \alpha \leq \infty$. This means that the method is *unconditionally stable!*

The above result seems too good to be true, since we have found an unconditionally stable method using an *explicit* combination of Lagrangian interpolation polynomials.

> The price we have paid for this is the loss of *consistency* with the original PDE.

To prove this, we expand the terms in (7.29) in a Taylor series and reconstruct the partial differential equation we are actually solving as the mesh size becomes very small. For the time derivative we have

$$\frac{1}{2\Delta t}\left(u_j^{(n+1)} - u_j^{(n-1)}\right) = (\partial_t u)_j^{(n)} + \frac{1}{6}\Delta t^2(\partial_{ttt}u)_j^{(n)} + \cdots$$

and for the mixed time and space differences

$$\frac{u_{j-1}^{(n)} - u_j^{(n+1)} - u_j^{(n-1)} + u_{j+1}^{(n)}}{\Delta x^2}$$

$$= (\partial_{xx}u)_j^{(n)} - \left(\frac{\Delta t}{\Delta x}\right)^2(\partial_{tt}u)_j^{(n)} + \frac{1}{12}\Delta x^2(\partial_{xxxx}u)_j^{(n)}$$

$$- \frac{1}{12}\Delta t^2\left(\frac{\Delta t}{\Delta x}\right)^2(\partial_{tttt}u)_j^{(n)} + \cdots . \tag{7.30}$$

Replace the terms in (7.29) with the above expansions and take the limit as Δt, $\Delta x \to 0$. We find

$$\frac{\partial u}{\partial t} = \nu\frac{\partial^2 u}{\partial x^2} - \nu r^2\frac{\partial^2 u}{\partial t^2} , \tag{7.31}$$

where

$$r \equiv \frac{\Delta t}{\Delta x} .$$

Equation (7.27) is *parabolic*. Equation (7.31) is *hyperbolic*. Thus if $\Delta t \to 0$ and $\Delta x \to 0$ in such a way that $\Delta t/\Delta x$ remains constant, the equation we actually solve by the method in (7.29) is a wave equation, not a diffusion equation. In such a case (7.29) *is not uniformly consistent with the equation we set out to solve* even for vanishingly small step sizes. If $\Delta t \to 0$ and $\Delta x \to 0$ in such a way that $\Delta t/\Delta x^2$ remains constant, then r^2 is $O(\Delta t)$, and the method is consistent. However, this leads to a constraint on the time step which is just as severe as that imposed by the stability condition of the second-order Runge–Kutta method shown in the table.

Exercises

7.1 Consider the ODE

$$u' = \frac{du}{dt} = Au + f$$

with

$$A = \begin{bmatrix} -10 & -0.1 & -0.1 \\ 1 & -1 & 1 \\ 10 & 1 & -1 \end{bmatrix} , \quad f = \begin{bmatrix} -1 \\ 0 \\ 0 \end{bmatrix} .$$

(a) Find the eigenvalues of A using a numerical package. What is the steady-state solution? How does the ODE solution behave in time?

(b) Write a code to integrate from the initial condition $u(0) = [1, 1, 1]^{\mathrm{T}}$ using the explicit Euler, implicit Euler, and MacCormack methods. In all three cases, use $h = 0.1$ for 1000 time steps, $h = 0.2$ for 500 time steps, $h = 0.4$ for 250 time steps and $h = 1.0$ for 100 time steps. Compare the computed solution with the exact steady solution.

(c) Using the λ–σ relations for these three methods, what are the expected bounds on h for stability? Are your results consistent with these bounds?

7.2 Perform σ-root traces for convection for the AB2 and RK2 methods.

(a) Compute a table of the numerical values of the σ-roots of the 2nd-order Adams–Bashforth method when $\lambda = i$. Take h in intervals of 0.05 from 0 to 0.80 and compute the absolute values of the roots to at least six places.

(b) Plot the trace of the roots in the complex σ-plane and draw the unit circle on the same plot.

(c) Repeat the above for the RK2 method.

7.3 When applied to the linear convection equation, the widely known Lax–Wendroff method gives:

$$u_j^{n+1} = u_j^n - \frac{1}{2}\frac{ah}{\Delta x}(u_{j+1}^n - u_{j-1}^n) + \frac{1}{2}\left(\frac{ah}{\Delta x}\right)^2 (u_{j+1}^n - 2u_j^n + u_{j-1}^n) .$$

Using Fourier stability analysis, find the range of $C_n = |ah/\Delta x|$ (known as the Courant number) for which the method is stable.

7.4 Determine and plot the stability contours for the Adams–Bashforth methods of order 1 through 4. Compare with the Runge–Kutta methods of order 1 through 4.

7.5 Recall the compact centered 3-point approximation for a first derivative

$$(\delta_x u)_{j-1} + 4(\delta_x u)_j + (\delta_x u)_{j+1} = \frac{3}{\Delta x}(u_{j+1} - u_{j-1}) \ .$$

By replacing the spatial index j by the temporal index n, obtain a time-marching method using this formula. What order is the method? Is it explicit or implicit? Is it a two-step LMM? If so, to what values of ξ, θ, and ϕ (in (6.59)) does it correspond? Derive the λ–σ relation for the method. Is it A-stable, A_0-stable, or I-stable?

7.6 Write the ODE system obtained by applying the 2nd-order centered difference approximation to the spatial derivative in the model diffusion equation with periodic boundary conditions. Using Appendix B.4, find the eigenvalues of the spatial operator matrix. Given that the λ–σ relation for the 2nd-order Adams–Bashforth method is

$$\sigma^2 - (1 + 3\lambda h/2)\sigma + \lambda h/2 = 0$$

show that the maximum stable value of $|\lambda h|$ for real negative λ, i.e. the point where the stability contour intersects the negative real axis, is obtained with $\lambda h = -1$. Using the eigenvalues of the spatial operator matrix, find the maximum stable time step for the combination of 2nd-order centered differences and the 2nd-order Adams–Bashforth method applied to the model diffusion equation. Repeat using Fourier analysis.

7.7 Consider the following PDE:

$$\frac{\partial u}{\partial t} = i\frac{\partial^2 u}{\partial x^2} \ ,$$

where $i = \sqrt{-1}$. Which explicit time-marching methods would be suitable for integrating this PDE, if 2nd-order centered differencing is used for the spatial differences and the boundary conditions are periodic? Find the stability condition for one of these methods.

7.8 Using Fourier analysis, analyze the stability of first-order backward differencing coupled with explicit Euler time marching applied to the linear convection equation with positive a. Find the maximum Courant number for stability.

7.9 Consider the linear convection equation with a positive wave speed as in Exercise 7.8. Apply a Dirichlet boundary condition at the left boundary. No boundary condition is permitted at the right boundary. Write the system of ODE's which results from first-order backward spatial differencing in matrix-vector form. Using Appendix B.1, find the λ-eigenvalues. Write the OΔE which results from the application of explicit Euler time marching in matrix-vector form, i.e.

$$\vec{u}_{n+1} = C\vec{u}_n - \vec{g}_n \ .$$

Write C in banded matrix notation and give the entries of $\vec{g}$. Using the λ–σ relation for the explicit Euler method, find the σ-eigenvalues. Based on these, what is the maximum Courant number allowed for asymptotic stability? Explain why this differs from the answer to Exercise 7.8. Hint: is C normal?

8. Choosing a Time-Marching Method

In this chapter, we discuss considerations involved in selecting a time-marching method for a specific application. Examples are given showing how time-marching methods can be compared in a given context. An important concept underlying much of this discussion is *stiffness*, which is defined in the next section.

8.1 Stiffness Definition for ODE's

8.1.1 Relation to λ-Eigenvalues

The introduction of the concept referred to as "stiffness" comes about from the numerical analysis of mathematical models constructed to simulate dynamic phenomena containing widely different time scales. Definitions given in the literature are not unique, but fortunately we now have the background material to construct a definition which is entirely sufficient for our purposes.

We start with the assumption that our CFD problem is modeled with sufficient accuracy by a coupled set of ODE's producing an A matrix typified by (7.1). Any definition of stiffness requires a *coupled* system with at least two eigenvalues, and the decision to use some numerical time-marching or iterative method to solve it. The difference between the dynamic scales in physical space is represented by the difference in the magnitude of the eigenvalues in eigenspace. In the following discussion, we concentrate on the transient part of the solution. The forcing function may also be time varying, in which case it would also have a time scale. However, we assume that this scale would be adequately resolved by the chosen time-marching method, and, since this part of the ODE has no effect on the numerical stability of the homogeneous part, we exclude the forcing function from further discussion in this section.

Consider now the form of the exact solution of a system of ODE's with a complete eigensystem. This is given by (6.27), and its solution using a one-root, time-marching method is represented by (6.28). For a given time step, the time integration is an approximation in eigenspace that is different for every eigenvector $\vec{x}_m$. In many numerical applications, the eigenvectors associated with the small $|\lambda_m|$ are well resolved, and those associated with the large $|\lambda_m|$ are resolved much less accurately, if at all. The situation is

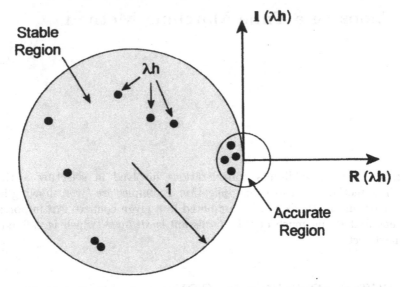

Fig. 8.1. Stable and accurate regions for the explicit Euler method

represented in the complex λh plane in Figure 8.1. In this figure, the time step has been chosen so that time accuracy is given to the eigenvectors associated with the eigenvalues lying in the small circle and stability without time accuracy is given to those associated with the eigenvalues lying outside of the small circle but still inside the large circle.

> The whole concept of stiffness in CFD arises from the fact that we often do not need the *time resolution* of eigenvectors associated with the large $|\lambda_m|$ in the transient solution, although these eigenvectors must remain coupled into the system to maintain a high accuracy of the *spatial resolution*.

8.1.2 Driving and Parasitic Eigenvalues

For the above reason it is convenient to subdivide the transient solution into two parts. First we order the eigenvalues by their magnitudes, thus

$$|\lambda_1| \le |\lambda_2| \le \cdots \le |\lambda_M| \ . \tag{8.1}$$

Then we write

$$\begin{matrix} \text{transient} \\ \text{solution} \end{matrix} = \underbrace{\sum_{m=1}^{p} c_m e^{\lambda_m t}\, \vec{x}_m}_{\text{driving}} + \underbrace{\sum_{m=p+1}^{M} c_m e^{\lambda_m t}\, \vec{x}_m}_{\text{parasitic}} \ . \tag{8.2}$$

This concept is crucial to our discussion. Rephrased, it states that we can separate our eigenvalue spectrum into two groups: one $[\lambda_1 \to \lambda_p]$ called the driving eigenvalues (our choice of a time-step and marching method must accurately approximate the time variation of the eigenvectors associated with these), and the other, $[\lambda_{p+1} \to \lambda_M]$, called the parasitic eigenvalues (no time accuracy whatsoever is required for the eigenvectors associated with these, but their presence must not contaminate the accuracy of the complete solution). Unfortunately, we find that, although time accuracy requirements are dictated by the driving eigenvalues, numerical stability requirements are dictated by the parasitic ones.

8.1.3 Stiffness Classifications

The following definitions are somewhat useful. An inherently stable set of ODE's is stiff if

$$|\lambda_p| \ll |\lambda_M| \ .$$

In particular we define the ratio

$$C_r = |\lambda_M| / |\lambda_p|$$

and form the categories

Mildly-stiff	$C_r < 10^2$
Strongly-stiff	$10^3 < C_r < 10^5$
Extremely-stiff	$10^6 < C_r < 10^8$
Pathologically-stiff	$10^9 < C_r$

It should be mentioned that the gaps in the stiff category definitions are intentional because the bounds are arbitrary. It is important to notice that these definitions make no distinction between real, complex, and imaginary eigenvalues.

8.2 Relation of Stiffness to Space Mesh Size

Many flow fields are characterized by a few regions having high spatial gradients of the dependent variables and other domains having relatively low gradient phenomena. As a result it is quite common to cluster mesh points in certain regions of space and spread them out otherwise. Examples of where this clustering might occur are at a shock wave, near an airfoil leading or trailing edge, and in a boundary layer.

One quickly finds that this grid clustering can strongly affect the eigensystem of the resulting A matrix. In order to demonstrate this, let us examine the eigensystems of the model problems given in Section 4.3.2. The simplest example to discuss relates to the model diffusion equation. In this case the eigenvalues are all real negative numbers that automatically obey the ordering given in (8.1). Consider the case when *all* of the eigenvalues are parasitic, i.e. we are interested only in the converged steady-state solution. Under these conditions, the stiffness is determined by the ratio λ_M/λ_1. A simple calculation shows that

$$\lambda_1 = -\frac{4\nu}{\Delta x^2} \sin^2\left(\frac{\pi}{2(M+1)}\right) \approx -\left(\frac{4\nu}{\Delta x^2}\right)\left(\frac{\Delta x}{2}\right)^2 = -\nu$$

$$\lambda_M \approx -\frac{4\nu}{\Delta x^2} \sin^2\left(\frac{\pi}{2}\right) = -\frac{4\nu}{\Delta x^2}$$

and the ratio is

$$\lambda_M/\lambda_1 \approx \frac{4}{\Delta x^2} = 4\left(\frac{M+1}{\pi}\right)^2 .$$

The most important information found from this example is the fact that the stiffness of the transient solution is directly related to the grid spacing. Furthermore, in diffusion problems this stiffness is proportional to the reciprocal of the space mesh size *squared*. For a mesh size $M = 40$, this ratio is about 680. Even for a mesh of this moderate size, the problem is already approaching the category of strongly-stiff.

For the biconvection model, a similar analysis shows that

$$|\lambda_M|/|\lambda_1| \approx \frac{1}{\Delta x} .$$

Here the stiffness parameter is still space-mesh dependent, but much less so than for diffusion-dominated problems.

We see that in both cases we are faced with the rather annoying fact that the more we try to increase the resolution of our spatial gradients, the stiffer our equations tend to become. Typical CFD problems without chemistry vary between the mildly and strongly-stiff categories, and are *greatly* affected by the resolution of a boundary layer since it is a diffusion process. Our brief analysis has been limited to equispaced problems, but in general the stiffness of CFD problems is proportional to the mesh intervals in the manner shown above where *the critical interval is the smallest one in the physical domain.*

8.3 Practical Considerations for Comparing Methods

We have presented relatively simple and reliable measures of stability and both the local and global accuracy of time-marching methods. Since there are an endless number of these methods to choose from, one can wonder how this information is to be used to pick a "best" choice for a particular problem. There is no unique answer to such a question. For example, it is, among other things, highly dependent upon the speed, capacity, and architecture of the available computer, and technology influencing this is undergoing rapid and dramatic changes as this is being written. Nevertheless, if certain ground rules are agreed upon, relevant conclusions can be reached. Let us now examine some ground rules that might be appropriate. It should then be clear how the analysis can be extended to other cases.

Let us consider the problem of measuring the efficiency of a time–marching method for computing, over a fixed interval of time, an accurate transient solution of a coupled set of ODE's. The length of the time interval, T, and the accuracy required of the solution are dictated by the physics of the particular problem involved. For example, in calculating the amount of turbulence in a homogeneous flow, the time interval would be that required to extract a reliable statistical sample, and the accuracy would be related to how much the energy of certain harmonics would be permitted to distort from a given level. Such a computation we refer to as an *event*.

The appropriate error measures to be used in comparing methods for calculating an event are the *global* ones, Er_a, Er_λ and Er_ω, discussed in Section 6.6.5, rather than the local ones, er_λ, er_a, and er_p, discussed earlier.

The actual form of the coupled ODE's that are produced by the semi-discrete approach is

$$\frac{d\vec{u}}{dt} = \vec{F}(\vec{u}, t) \ .$$

At every time step we must evaluate the function $\vec{F}(\vec{u}, t)$ at least once. This function is usually nonlinear, and its computation *usually consumes the major portion of the computer time required to make the simulation.* We refer to a single calculation of the vector $\vec{F}(\vec{u}, t)$ as a *function evaluation* and denote the total number of such evaluations by F_{ev}.

8.4 Comparing the Efficiency of Explicit Methods

8.4.1 Imposed Constraints

As mentioned above, the efficiency of methods can be compared only if one accepts a set of limiting constraints within which the comparisons are carried out. The following assumptions bound the considerations made in this section:

1. The time-march method is explicit.
2. Implications of computer storage capacity and access time are ignored. In some contexts, this can be an important consideration.
3. The calculation is to be time-accurate, must simulate an entire event which takes a total time T, and must use a constant time step size, h, so that

$$T = Nh \ ,$$

where N is the total number of time steps.

8.4.2 An Example Involving Diffusion

Let the event be the numerical solution of

$$\frac{du}{dt} = -u \qquad (8.3)$$

from $t = 0$ to $T = -\ln(0.25)$ with $u(0) = 1$. Equation (8.3) is obtained from our representative ODE with $\lambda = -1$, $a = 0$. Since the exact solution is $u(t) = u(0)e^{-t}$, this makes the exact value of u at the end of the event equal to 0.25, i.e. $u(T) = 0.25$. To the constraints imposed above, let us set the additional requirement

• The error in u at the end of the event, i.e. the *global* error, must be $< 0.5\%$.

We judge the most efficient method to be the one that satisfies these conditions and has the smallest number of function evaluations, F_{ev}. Three methods are compared – explicit Euler, AB2, and RK4.

First of all, the allowable error constraint means that the global error in the amplitude (see (6.48)) must have the property

$$\left| \frac{Er_\lambda}{e^{\lambda T}} \right| < 0.005 \ .$$

Then, since $h = T/N = -\ln(0.25)/N$, it follows that

$$\left| 1 - \{\sigma_1[\ln(0.25)/N]\}^N/0.25 \right| < 0.005 \ ,$$

where σ_1 is found from the characteristic polynomials given in Table 7.1. The results shown in Table 8.1 were computed using a simple iterative procedure.

Table 8.1. Comparison of time-marching methods for a simple diffusion problem

Method	N	h	σ_1	F_{ev}	Er_λ	
Euler	193	.00718	.99282	193	.001248	worst
AB2	16	.0866	.9172	16	.001137	
RK4	2	.6931	.5012	8	.001195	best

In this example we see that, for a given global accuracy, the method with the highest local accuracy is the most efficient on the basis of the expense in terms of function evaluations. Thus the second-order Adams–Bashforth method is much better than the first-order Euler method, and the fourth-order Runge–Kutta method is the best of all. The main purpose of this exercise is to show the (usually) great superiority of second-order over first-order time-marching methods.

8.4.3 An Example Involving Periodic Convection

Let us use as a basis for this example the study of homogeneous turbulence simulated by the numerical solution of the incompressible Navier–Stokes equations inside a cube with periodic boundary conditions on all sides. In this numerical experiment the function evaluations contribute overwhelmingly to the CPU time, and the number of these evaluations must be kept to an absolute minimum because of the magnitude of the problem. On the other hand, a complete event must be established in order to obtain meaningful statistical samples which are the essence of the solution. In this case, in addition to the constraints given in Section 8.4.1, we add the following:

- The number of evaluations of $\vec{F}(\vec{u}, t)$ is fixed.

Under these conditions a method is judged as best when it has the highest global accuracy for resolving eigenvectors with imaginary eigenvalues. The above constraint has led to the invention of schemes that omit the function evaluation in the corrector step of a predictor–corrector combination, leading to the so-called incomplete predictor–corrector methods. The presumption is, of course, that more efficient methods will result from the omission of the second function evaluation. An example is the method of Gazdag, given in Section 6.8. Basically this is composed of an AB2 predictor and a trapezoidal corrector. However, the derivative of the fundamental family is never found, so there is only one evaluation required to complete each cycle. The λ–σ relation for the method is shown as entry 10 in Table 7.1.

In order to discuss our comparisons, we introduce the following definitions:

- Let a k-evaluation method be defined as one that requires k evaluations of $\vec{F}(\vec{u}, t)$ to advance one step using that method's time interval, h.
- Let K represent the total number of allowable function evaluations.
- Let h_1 be the time interval advanced in one step of a one-evaluation method.

The Gazdag, leapfrog, and AB2 schemes are all 1-evaluation methods. The second- and fourth-order RK methods are 2- and 4-evaluation methods, respectively. For a 1-evaluation method, the total number of time steps, N, and the number of evaluations, K, are the same, one evaluation being used for each step, so that for these methods $h = h_1$. For a 2-evaluation method, $N = K/2$ since two evaluations are used for each step. However, in this case, in order to arrive at the same time T after K evaluations, the time step must be twice that of a one-evaluation method so $h = 2h_1$. For a 4-evaluation method the time interval must be $h = 4h_1$, etc. Notice that as k increases, the time span required for one application of the method increases. *However, notice also that as k increases, the power to which σ_1 is raised to arrive at the final destination decreases.* This is the key to the true comparison of time-march methods for this type of problem.

		0							T	u_N
$k = 1$	|	•	•	•	•	•	•	•	|	$[\sigma(\lambda h_1)]^8$
$k = 2$	|	$2h_1$	•		•		•		|	$[\sigma(2\lambda h_1)]^4$
$k = 4$	|	$4h_1$			•				|	$[\sigma(4\lambda h_1)]^2$

Step sizes and powers of σ for k-evaluation methods used to get to the same value of T if 8 evaluations are allowed.

In general, after K evaluations, the global amplitude and phase error for k-evaluation methods applied to systems with pure imaginary λ-roots can be written[1]

$$Er_a = 1 - |\sigma_1(ik\omega h_1)|^{K/k} \tag{8.4}$$

$$Er_\omega = \omega T - \frac{K}{k} \tan^{-1}\left[\frac{[\sigma_1(ik\omega h_1)]_{\text{imaginary}}}{[\sigma_1(ik\omega h_1)]_{\text{real}}}\right]. \tag{8.5}$$

Consider a convection-dominated event for which the function evaluation is very time consuming. We idealize to the case where $\lambda = i\omega$ and set ω equal to one. The event must proceed to the time $t = T = 10$. We consider two maximum evaluation limits $K = 50$ and $K = 100$ and choose from four possible methods, leapfrog, AB2, Gazdag, and RK4. The first three of these are one-evaluation methods and the last one is a four-evaluation method. It is not difficult to show that on the basis of local error (made in a single step) the Gazdag method is superior to the RK4 method in both amplitude and phase. For example, for $\omega h = 0.2$ the Gazdag method produces a $|\sigma_1| = 0.9992276$, whereas for $\omega h = 0.8$ (which must be used to keep the number of evaluations the same) the RK4 method produces a $|\sigma_1| = 0.998324$. However, we are making our comparisons on the basis of global error for a fixed number of evaluations.

First of all, we see that for a one-evaluation method $h_1 = T/K$. Using this, and the fact that $\omega = 1$, we find, by some rather simple calculations[2] made using (8.4) and (8.5), the results shown in Table 8.2. Notice that to find global error the Gazdag root must be raised to the power of 50, while the RK4 root is raised only to the power of 50/4. On the basis of *global error*, the Gazdag method is not superior to RK4 in either amplitude or phase, although, in terms of phase error (for which it was designed) it is superior to the other two methods shown.

---

[1] See (6.49) and (6.50).

[2] The σ_1-root for the Gazdag method can be found using a numerical root finding routine to trace the three roots in the σ-plane (see Figure 7.2e).

Table 8.2 Comparison of global amplitude and phase errors for
four methods

	K	leapfrog	AB2	Gazdag	RK4
$\omega h_1 = 0.1$	100	1.0	1.003	0.995	0.999
$\omega h_1 = 0.2$	50	1.0	1.022	0.962	0.979

a. Amplitude, exact $= 1.0$.

	K	leapfrog	AB2	Gazdag	RK4
$\omega h_1 = 0.1$	100	-0.96	-2.4	0.45	0.12
$\omega h_1 = 0.2$	50	-3.8	-9.8	1.5	1.5

b. Phase error in degrees.

Using analysis such as this (and also considering the stability boundaries) the RK4 method is recommended as a basic first choice for any explicit time-accurate calculation of a convection-dominated problem.

8.5 Coping with Stiffness

8.5.1 Explicit Methods

The ability of a numerical method to cope with stiffness can be illustrated quite nicely in the complex λh plane. A good example of the concept is produced by studying the Euler method applied to the representative equation. The transient solution is $u_n = (1 + \lambda h)^n$ and the trace of the complex value of λh which makes $|1 + \lambda h| = 1$ gives the whole story. In this case the trace forms a circle of unit radius centered at $(-1, 0)$ as shown in Figure 8.1. If h is chosen so that all λh in the ODE eigensystem fall inside this circle, the integration will be numerically stable. Also shown by the small circle centered at the origin is the region of Taylor series accuracy. If some λh fall outside the small circle but stay within the stable region, these λh are stiff, but stable. We have defined these λh as parasitic eigenvalues. Stability boundaries for some explicit methods are shown in Figs. 7.4 and 7.5.

For a specific example, consider the mildly-stiff system composed of a coupled two-equation set having the two eigenvalues $\lambda_1 = -100$ and $\lambda_2 = -1$. If uncoupled and evaluated in wave space, the time histories of the two solutions would appear as a rapidly decaying function in one case, and a relatively slowly decaying function in the other. Analytical evaluation of the time histories poses no problem since e^{-100t} quickly becomes very small and can be neglected in the expressions when time becomes large. Numerical evaluation is altogether different. Numerical solutions, of course, depend upon

$[\sigma(\lambda_m h)]^n$ and no $|\sigma_m|$ can exceed one for any λ_m in the coupled system, or else the process is numerically unstable.

Let us choose the simple explicit Euler method for the time march. The coupled equations in real space are represented by

$$u_1(n) = c_1(1 - 100h)^n x_{11} + c_2(1 - h)^n x_{12} + (P.S.)_1$$
$$u_2(n) = c_1(1 - 100h)^n x_{21} + c_2(1 - h)^n x_{22} + (P.S.)_2 \ . \qquad (8.6)$$

We will assume that our accuracy requirements are such that sufficient accuracy is obtained as long as $|\lambda h| \le 0.1$. This defines a time step limit based on *accuracy* considerations of $h = 0.001$ for λ_1 and $h = 0.1$ for λ_2. The time step limit based on *stability*, which is determined from λ_1, is $h = 0.02$. We will also assume that $c_1 = c_2 = 1$ and that an amplitude less than 0.001 is negligible. We first run 66 time steps with $h = 0.001$ in order to resolve the λ_1 term. With this time step the λ_2 term is resolved exceedingly well. After 66 steps, the amplitude of the λ_1 term (i.e. $(1 - 100h)^n$) is less than 0.001 and that of the λ_2 term (i.e. $(1 - h)^n$) is 0.9361. Hence the λ_1 term can now be considered negligible. To drive the $(1-h)^n$ term to zero (i.e. below 0.001), we would like to change the step size to $h = 0.1$ and continue. We would then have a well resolved answer to the problem throughout the entire relevant time interval. However, this is not possible because of the coupled presence of $(1 - 100h)^n$, which in just ten steps at $h = 0.1$ amplifies those terms by $\approx 10^9$, far outweighing the initial decrease obtained with the smaller time step. In fact, with $h = 0.02$, the maximum step size that can be taken in order to maintain stability, about 339 time steps have to be computed in order to drive e^{-t} to below 0.001. Thus the total simulation requires 405 time steps.

8.5.2 Implicit Methods

Now let us re-examine the problem that produced (8.6) but this time using an unconditionally stable implicit method for the time march. We choose the trapezoidal method. Its behavior in the λh plane is shown in Figure 7.3b. Since this is also a one-root method, we simply replace the Euler σ with the trapezoidal one and analyze the result. It follows that the final numerical solution to the ODE is now represented in real space by

$$u_1(n) = c_1\left(\frac{1 - 50h}{1 + 50h}\right)^n x_{11} + c_2\left(\frac{1 - 0.5h}{1 + 0.5h}\right)^n x_{12} + (P.S.)_1$$
$$u_2(n) = c_1\left(\frac{1 - 50h}{1 + 50h}\right)^n x_{21} + c_2\left(\frac{1 - 0.5h}{1 + 0.5h}\right)^n x_{22} + (P.S.)_2 \ . \qquad (8.7)$$

In order to resolve the initial transient of the term e^{-100t}, we need to use a step size of about $h = 0.001$. This is the same step size used in applying the explicit Euler method because here accuracy is the only consideration and a very small step size must be chosen to get the desired resolution. (It is true that for the same accuracy we could in this case use a larger step size because this is a second-order method, but that is not the point of this exercise.) After 70 time steps, the λ_1 term has an amplitude less than 0.001 and can be neglected. Now with the implicit method we can proceed to calculate the remaining part of the event using our desired step size $h = 0.1$ without any problem of instability, with 69 steps required to reduce the amplitude of the second term to below 0.001. In both intervals the desired solution is second-order accurate and well resolved. It is true that in the final 69 steps one σ-root is $[1 - 50(0.1)]/[1 + 50(0.1)] = -0.666\cdots$, and this has no physical meaning whatsoever. However, its influence on the coupled solution is negligible at the end of the first 70 steps, and, since $(0.666\cdots)^n < 1$, its influence in the remaining 69 steps is even less. Actually, although this root is one of the principal roots in the system, its behavior for $t > 0.07$ is identical to that of a stable spurious root. The total simulation requires 139 time steps.

8.5.3 A Perspective

It is important to retain a proper perspective on problems represented by the above example. It is clear that an unconditionally stable method can always be called upon to solve stiff problems with a minimum number of time steps. In the example, the conditionally stable Euler method required 405 time steps, as compared to about 139 for the trapezoidal method, about three times as many. However, the Euler method is extremely easy to program and requires very little arithmetic per step. For preliminary investigations it is often the best method to use for mildly-stiff diffusion dominated problems. For refined investigations of such problems an explicit method of second order or higher, such as Adams–Bashforth or Runge–Kutta methods, is recommended. These explicit methods can be considered as effective mildly stiff-stable methods. However, it should be clear that as the degree of stiffness of the problem increases, the advantage begins to tilt towards implicit methods, as the reduced number of time steps begins to outweigh the increased cost per time step. The reader can repeat the above example with $\lambda_1 = -10,000$, $\lambda_2 = -1$, which is in the strongly-stiff category.

There is yet another technique for coping with certain stiff systems in fluid dynamics applications. This is known as the *multigrid* method. It has enjoyed remarkable success in many practical problems; however, we need an introduction to the theory of relaxation before it can be presented.

8.6 Steady Problems

In Chapter 6, we wrote the OΔE solution in terms of the principal and spurious roots as follows:

$$u_n = c_{11}(\sigma_1)_1^n \vec{x}_1 + \cdots + c_{m1}(\sigma_m)_1^n \vec{x}_m + \cdots + c_{M1}(\sigma_M)_1^n \vec{x}_M + P.S.$$
$$+ c_{12}(\sigma_1)_2^n \vec{x}_1 + \cdots + c_{m2}(\sigma_m)_2^n \vec{x}_m + \cdots + c_{M2}(\sigma_M)_2^n \vec{x}_M$$
$$+ c_{13}(\sigma_1)_3^n \vec{x}_1 + \cdots + c_{m3}(\sigma_m)_3^n \vec{x}_m + \cdots + c_{M3}(\sigma_M)_3^n \vec{x}_M$$
$$+ \text{etc., if there are more spurious roots.} \tag{8.8}$$

When solving a *steady* problem, we have no interest whatsoever in the transient portion of the solution. Our sole goal is to eliminate it as quickly as possible. Therefore, the choice of a time-marching method for a steady problem is similar to that for a stiff problem, the difference being that the order of accuracy is irrelevant. Hence the explicit Euler method is a candidate for steady diffusion-dominated problems, and the fourth-order Runge–Kutta method is a candidate for steady convection-dominated problems, because of their stability properties. Among implicit methods, the implicit Euler method is the obvious choice for steady problems.

When we seek only the steady solution, all of the eigenvalues can be considered to be parasitic. Referring to Figure 8.1, none of the eigenvalues are required to fall in the accurate region of the time-marching method. Therefore the time step can be chosen to eliminate the transient as quickly as possible with no regard for time accuracy. For example, when using the implicit Euler method with local time linearization, (6.96), one would like to take the limit $h \to \infty$, which leads to Newton's method, (6.98). However, a finite time step may be required until the solution is somewhat close to the steady solution.

Exercises

8.1 Repeat the time-march comparisons for diffusion (Section 8.4.2) and periodic convection (Section 8.4.3) using 2nd- and 3rd-order Runge–Kutta methods.

8.2 Repeat the time-march comparisons for diffusion (Section 8.4.2) and periodic convection (Section 8.4.3) using the 3rd- and 4th-order Adams–Bashforth methods. Considering the stability bounds for these methods (see Exercise 7.4) as well as their memory requirements, compare and contrast them with the 3rd- and 4th-order Runge–Kutta methods.

8.3 Consider the diffusion equation (with $\nu = 1$) discretized using 2nd-order central differences on a grid with ten (interior) points. Find and plot the eigenvalues and the corresponding modified wavenumbers. If we use the explicit Euler time-marching method what is the maximum allowable time step if all but the first two eigenvectors are considered parasitic? Assume that sufficient accuracy is obtained as long as $|\lambda h| \leq 0.1$. What is the maximum allowable time step if all but the first eigenvector are considered parasitic?

9. Relaxation Methods

In the past three chapters, we developed a methodology for designing, analyzing, and choosing time-marching methods. These methods can be used to compute the time-accurate solution to linear and nonlinear systems of ODE's in the general form

$$\frac{d\vec{u}}{dt} = \vec{F}(\vec{u}, t) ,$$

(9.1)

which arise after spatial discretization of a PDE. Alternatively, they can be used to solve for the steady solution of (9.1), which satisfies the following coupled system of nonlinear algebraic equations:

$$\vec{F}(\vec{u}) = 0 .$$

(9.2)

In the latter case, the unsteady equations are integrated until the solution converges to a steady solution. The same approach permits a time-marching method to be used to solve a linear system of algebraic equations in the form

$$A\vec{x} = \vec{b} .$$

(9.3)

To solve this system using a time-marching method, a time derivative is introduced as follows

$$\frac{d\vec{x}}{dt} = A\vec{x} - \vec{b}$$

(9.4)

and the system is integrated in time until the transient has decayed to a sufficiently low level. Following a time-dependent path to steady state is possible only if all of the eigenvalues of the matrix A (or $-A$) have real parts lying in the left half-plane. Although the solution $\vec{x} = A^{-1}\vec{b}$ exists as long as A is nonsingular, the ODE given by (9.4) has a stable steady solution only if A meets the above condition.

The common feature of all time-marching methods is that they are at least first-order accurate. In this chapter, we consider iterative methods which are not time-accurate at all. Such methods are known as *relaxation methods*. While they are applicable to coupled systems of nonlinear algebraic equations in the form of (9.2), our analysis will focus on their application to large sparse linear systems of equations in the form

$$A_b \vec{u} - \vec{f}_b = 0 \ , \qquad\qquad (9.5)$$

where A_b is nonsingular, and the use of the subscript b will become clear shortly. Such systems of equations arise, for example, at each time step of an implicit time-marching method or at each iteration of Newton's method. Using an iterative method, we seek to obtain rapidly a solution which is arbitrarily close to the exact solution of (9.5), which is given by

$$\vec{u}_\infty = A_b^{-1} \vec{f}_b \ . \qquad\qquad (9.6)$$

9.1 Formulation of the Model Problem

9.1.1 Preconditioning the Basic Matrix

It is standard practice in applying relaxation procedures to *precondition* the basic equation. This preconditioning has the effect of multiplying (9.5) from the left by some nonsingular matrix. In the simplest possible case, the conditioning matrix is a diagonal matrix composed of a constant $D(b)$. If we designate the conditioning matrix by C, the problem becomes one of solving for $\vec{u}$ in

$$CA_b \vec{u} - C\vec{f}_b = 0 \ . \qquad\qquad (9.7)$$

Notice that the solution of (9.7) is

$$\vec{u} = [CA_b]^{-1} C\vec{f}_b = A_b^{-1} C^{-1} C\vec{f}_b = A_b^{-1} \vec{f}_b \ , \qquad\qquad (9.8)$$

which is identical to the solution of (9.5), provided C^{-1} exists.

In the following, we will see that our approach to the iterative solution of (9.7) depends crucially on the eigenvalue and eigenvector structure of the matrix CA_b, and, equally important, does not depend at all on the eigensystem of the basic matrix A_b. For example, there are well-known techniques for accelerating relaxation schemes if the eigenvalues of CA_b are all real and of the same sign. To use these schemes, the conditioning matrix C must be chosen such that this requirement is satisfied. A choice of C which ensures this condition is the negative transpose of A_b.

For example, consider a spatial discretization of the linear convection equation using centered differences with a Dirichlet condition on the left side and no constraint on the right side. Using a first-order backward difference on the right side (as in Section 3.6), this leads to the approximation

$$\delta_x \vec{u} = \frac{1}{2\Delta x} \left\{ \begin{bmatrix} 0 & 1 & & & \\ -1 & 0 & 1 & & \\ & -1 & 0 & 1 & \\ & & -1 & 0 & 1 \\ & & & -2 & 2 \end{bmatrix} \vec{u} + \begin{bmatrix} -u_a \\ 0 \\ 0 \\ 0 \\ 0 \end{bmatrix} \right\} . \qquad (9.9)$$

The matrix in (9.9) has eigenvalues whose imaginary parts are much larger than their real parts. It can first be conditioned so that the modulus of each element is 1. This is accomplished using a diagonal preconditioning matrix

$$D = 2\Delta x \begin{bmatrix} 1 & 0 & 0 & 0 & 0 \\ 0 & 1 & 0 & 0 & 0 \\ 0 & 0 & 1 & 0 & 0 \\ 0 & 0 & 0 & 1 & 0 \\ 0 & 0 & 0 & 0 & \frac{1}{2} \end{bmatrix}, \tag{9.10}$$

which scales each row. We then further condition with multiplication by the negative transpose. The result is

$$A_2 = -A_1^T A_1$$

$$= \begin{bmatrix} 0 & 1 & & & \\ -1 & 0 & 1 & & \\ & -1 & 0 & 1 & \\ & & -1 & 0 & 1 \\ & & & -1 & -1 \end{bmatrix} \begin{bmatrix} 0 & 1 & & & \\ -1 & 0 & 1 & & \\ & -1 & 0 & 1 & \\ & & -1 & 0 & 1 \\ & & & -1 & 1 \end{bmatrix}$$

$$= \begin{bmatrix} -1 & 0 & 1 & & \\ 0 & -2 & 0 & 1 & \\ 1 & 0 & -2 & 0 & 1 \\ & 1 & 0 & -2 & 1 \\ & & 1 & 1 & -2 \end{bmatrix} . \tag{9.11}$$

If we define a permutation matrix P [1] and carry out the process $P^T[-A_1^T A_1]P$ (which just reorders the elements of A_1 and doesn't change the eigenvalues) we find

$$\begin{bmatrix} 0 & 1 & 0 & 0 & 0 \\ 0 & 0 & 0 & 1 & 0 \\ 0 & 0 & 0 & 0 & 1 \\ 0 & 0 & 1 & 0 & 0 \\ 1 & 0 & 0 & 0 & 0 \end{bmatrix} \begin{bmatrix} -1 & 0 & 1 & & \\ 0 & -2 & 0 & 1 & \\ 1 & 0 & -2 & 0 & 1 \\ & 1 & 0 & -2 & 1 \\ & & 1 & 1 & -2 \end{bmatrix} \begin{bmatrix} 0 & 0 & 0 & 0 & 1 \\ 1 & 0 & 0 & 0 & 0 \\ 0 & 0 & 0 & 1 & 0 \\ 0 & 1 & 0 & 0 & 0 \\ 0 & 0 & 1 & 0 & 0 \end{bmatrix}$$

$$= \begin{bmatrix} -2 & 1 & & & \\ 1 & -2 & 1 & & \\ & 1 & -2 & 1 & \\ & & 1 & -2 & 1 \\ & & & 1 & -1 \end{bmatrix} , \tag{9.12}$$

---

[1] A permutation matrix (defined as a matrix with exactly one 1 in each row and column and has the property that $P^T = P^{-1}$) just rearranges the rows and columns of a matrix.

which has all negative real eigenvalues, as given in Appendix B. Thus even when the basic matrix A_b has nearly imaginary eigenvalues, the conditioned matrix $-A_b^\mathrm{T} A_b$ is nevertheless symmetric negative definite (i.e. symmetric with negative real eigenvalues), and the classical relaxation methods can be applied. We do not necessarily recommend the use of $-A_b^\mathrm{T}$ as a preconditioner; we simply wish to show that a broad range of matrices can be preconditioned into a form suitable for our analysis.

9.1.2 The Model Equations

Preconditioning processes such as those described in the last section allow us to prepare our algebraic equations in advance so that certain eigenstructures are guaranteed. In the remainder of this chapter, we will thoroughly investigate some simple equations which model these structures. We will consider the preconditioned system of equations having the form

$$A\vec{\phi} - \vec{f} = 0 , \tag{9.13}$$

where A is symmetric negative definite.[2] The symbol for the dependent variable has been changed to ϕ as a reminder that the physics being modeled is no longer time accurate when we later deal with ODE formulations. Note that the solution of (9.13), $\vec{\phi} = A^{-1}\vec{f}$, is guaranteed to exist because A is nonsingular. In the notation of (9.5) and (9.7),

$$A = C A_b \quad \text{and} \quad \vec{f} = C\vec{f_b} . \tag{9.14}$$

The above was written to treat the general case. It is instructive in formulating the concepts to consider the special case given by the diffusion equation in one dimension with unit diffusion coefficient ν:

$$\frac{\partial u}{\partial t} = \frac{\partial^2 u}{\partial x^2} - g(x) . \tag{9.15}$$

The steady-state solution satisfies

$$\frac{\partial^2 u}{\partial x^2} = g(x) . \tag{9.16}$$

Introducing the three-point central differencing scheme for the second derivative with Dirichlet boundary conditions, we find

$$\frac{\mathrm{d}\vec{u}}{\mathrm{d}t} = \frac{1}{\Delta x^2} B(1, -2, 1)\vec{u} + (\vec{bc}) - \vec{g} , \tag{9.17}$$

---

[2] We use a symmetric negative definite matrix to simplify certain aspects of our analysis. Relaxation methods are applicable to more general matrices. The classical methods will usually converge if A_b is *diagonally dominant*, as defined in Appendix A.

where $(\vec{bc})$ contains the boundary conditions, and $\vec{g}$ contains the values of the source term at the grid nodes. In this case

$$A_b = \frac{1}{\Delta x^2} B(1, -2, 1)$$
$$\vec{f_b} = \vec{g} - (\vec{bc}) \ . \tag{9.18}$$

Choosing $C = \Delta x^2 I$, we obtain

$$B(1, -2, 1)\vec{\phi} = \vec{f} \ , \tag{9.19}$$

where $\vec{f} = \Delta x^2 \vec{f_b}$. If we consider a Dirichlet boundary condition on the left side and either a Dirichlet or a Neumann condition on the right side, then A has the form

$$A = B\left(1, \vec{b}, 1\right)$$
$$\vec{b} = [-2, -2, \cdots, -2, s]^{\mathrm{T}} \tag{9.20}$$
$$s = -2 \quad \text{or} \quad -1 \ .$$

Note that $s = -1$ is easily obtained from the matrix resulting from the Neumann boundary condition given in (3.24) using a diagonal conditioning matrix. A tremendous amount of insight to the basic features of relaxation is gained by an appropriate study of the one-dimensional case, and much of the remaining material is devoted to this case. We attempt to do this in such a way, however, that it is directly applicable to two- and three-dimensional problems.

9.2 Classical Relaxation

9.2.1 The Delta Form of an Iterative Scheme

We will consider relaxation methods which can be expressed in the following delta form:

$$H\left[\vec{\phi}_{n+1} - \vec{\phi}_n\right] = A\vec{\phi}_n - \vec{f} \ , \tag{9.21}$$

where H is some nonsingular matrix which depends upon the iterative method. The matrix H is independent of n for *stationary* methods and is a function of n for *nonstationary* ones. It is clear that H should lead to a system of equations which is easy to solve, or at least easier to solve than the original system. The iteration count is designated by the subscript n or the superscript (n). The converged solution is designated $\vec{\phi}_\infty$ so that

$$\vec{\phi}_\infty = A^{-1}\vec{f} \ . \tag{9.22}$$

9.2.2 The Converged Solution, the Residual, and the Error

Solving (9.21) for $\vec{\phi}_{n+1}$ gives

$$\vec{\phi}_{n+1} = [I + H^{-1}A]\vec{\phi}_n - H^{-1}\vec{f} = G\vec{\phi}_n - H^{-1}\vec{f} , \qquad (9.23)$$

where

$$G \equiv I + H^{-1}A . \qquad (9.24)$$

The *error* at the nth iteration is defined as

$$\vec{e}_n \equiv \vec{\phi}_n - \vec{\phi}_\infty = \vec{\phi}_n - A^{-1}\vec{f} , \qquad (9.25)$$

where $\vec{\phi}_\infty$ is defined in (9.22). The *residual* at the nth iteration is defined as

$$\vec{r}_n \equiv A\vec{\phi}_n - \vec{f} . \qquad (9.26)$$

Multiply (9.25) by A from the left, and use the definition in (9.26). There results the relation between the error and the residual

$$A\vec{e}_n - \vec{r}_n = 0 . \qquad (9.27)$$

Finally, it is not difficult to show that

$$\vec{e}_{n+1} = G\vec{e}_n . \qquad (9.28)$$

Consequently, G is referred to as the basic iteration matrix, and its eigenvalues, which we designate as σ_m, determine the convergence rate of a method.

In all of the above, we have considered only what are usually referred to as stationary processes, in which H is constant throughout the iterations. Non-stationary processes, in which H (and possibly C) is varied at each iteration, are discussed in Section 9.5.

9.2.3 The Classical Methods

Point Operator Schemes in One Dimension. Let us consider three classical relaxation procedures for our model equation

$$B(1, -2, 1)\vec{\phi} = \vec{f} \qquad (9.29)$$

as given in Section 9.1.2. The point-Jacobi method is expressed in point operator form for the one-dimensional case as

$$\phi_j^{(n+1)} = \frac{1}{2}\left[\phi_{j-1}^{(n)} + \phi_{j+1}^{(n)} - f_j\right] . \qquad (9.30)$$

This operator comes about by choosing the value of $\phi_j^{(n+1)}$ such that together with the *old* values of ϕ_{j-1} and ϕ_{j+1}, the jth row of (9.29) is satisfied. The Gauss–Seidel method is

$$\phi_j^{(n+1)} = \frac{1}{2}\left[\phi_{j-1}^{(n+1)} + \phi_{j+1}^{(n)} - f_j\right]. \tag{9.31}$$

This operator is a simple extension of the point-Jacobi method which uses the most recent update of ϕ_{j-1}. Hence the jth row of (9.29) is satisfied using the *new* values of ϕ_j and ϕ_{j-1} and the *old* value of ϕ_{j+1}. The method of successive overrelaxation (SOR) is based on the idea that if the correction produced by the Gauss–Seidel method tends to move the solution toward $\vec{\phi}_\infty$, then perhaps it would be better to move further in this direction. It is usually expressed in two steps as

$$\tilde{\phi}_j = \frac{1}{2}\left[\phi_{j-1}^{(n+1)} + \phi_{j+1}^{(n)} - f_j\right]$$

$$\phi_j^{(n+1)} = \phi_j^{(n)} + \omega\left[\tilde{\phi}_j - \phi_j^{(n)}\right], \tag{9.32}$$

where ω generally lies between 1 and 2, but it can also be written in the single line

$$\phi_j^{(n+1)} = \frac{\omega}{2}\phi_{j-1}^{(n+1)} + (1-\omega)\phi_j^{(n)} + \frac{\omega}{2}\phi_{j+1}^{(n)} - \frac{\omega}{2}f_j. \tag{9.33}$$

The General Form. The general form of the classical methods is obtained by splitting the matrix A in (9.13) into its diagonal, D, the portion of the matrix below the diagonal, L, and the portion above the diagonal, U, such that

$$A = L + D + U. \tag{9.34}$$

Then the point-Jacobi method is obtained with $H = -D$, which certainly meets the criterion that it is easy to solve. The Gauss–Seidel method is obtained with $H = -(L+D)$, which is also easy to solve, being lower triangular.

9.3 The ODE Approach to Classical Relaxation

9.3.1 The Ordinary Differential Equation Formulation

The particular type of delta form given by (9.21) leads to an interpretation of relaxation methods in terms of solution techniques for coupled first-order ODE's, about which we have already learned a great deal. One can easily see that (9.21) results from the application of the explicit Euler time-marching method (with $h = 1$) to the following system of ODE's:

$$H\frac{d\vec{\phi}}{dt} = A\vec{\phi} - \vec{f}. \tag{9.35}$$

This is equivalent to

$$\frac{d\vec{\phi}}{dt} = H^{-1}C\left[A_b\vec{\phi} - \vec{f_b}\right] = H^{-1}[A\vec{\phi} - \vec{f}] . \qquad (9.36)$$

In the special case where $H^{-1}A$ depends on neither $\vec{u}$ nor t, $H^{-1}\vec{f}$ is also independent of t, *and* the eigenvectors of $H^{-1}A$ are linearly independent, the solution can be written as

$$\vec{\phi} = \underbrace{c_1 e^{\lambda_1 t}\vec{x}_1 + \cdots + c_M e^{\lambda_M t}\vec{x}_M}_{\text{error}} + \vec{\phi}_\infty , \qquad (9.37)$$

where what is referred to in time-accurate analysis as the transient solution is now referred to in relaxation analysis as the error. It is clear that, if all of the eigenvalues of $H^{-1}A$ have negative real parts (which implies that $H^{-1}A$ is nonsingular), then the system of ODE's has a *steady-state* solution which is approached as $t \to \infty$, given by

$$\vec{\phi}_\infty = A^{-1}\vec{f} , \qquad (9.38)$$

which is the solution of (9.13). We see that the goal of a relaxation method is to remove the transient solution from the general solution in the most efficient way possible. The λ eigenvalues are fixed by the basic matrix A_b in (9.36), the preconditioning matrix C in (9.7), and the secondary conditioning matrix H in (9.35). The σ eigenvalues are fixed for a given λh by the choice of time-marching method. Throughout the remaining discussion we will refer to the independent variable t as "time", even though no true time accuracy is involved.

In a stationary method, H and C in (9.36) are independent of t, that is, they are not changed throughout the iteration process. The generalization of this in our approach is to make h, the "time" step, a constant for the entire iteration.

Suppose the explicit Euler method is used for the time integration. For this method $\sigma_m = 1 + \lambda_m h$. Hence the numerical solution after n steps of a stationary relaxation method can be expressed as (see (6.28))

$$\vec{\phi}_n = \underbrace{c_1\vec{x}_1(1 + \lambda_1 h)^n + \cdots + c_m\vec{x}_m(1 + \lambda_m h)^n + \cdots + c_M\vec{x}_M(1 + \lambda_M h)^n}_{\text{error}}$$

$$+ \vec{\phi}_\infty . \qquad (9.39)$$

The initial amplitudes of the eigenvectors are given by the magnitudes of the c_m. These are fixed by the initial guess. In general it is assumed that any or all of the eigenvectors could have been given an equally "bad" excitation by the initial guess, so that we must devise a way to remove them all from the general solution on an equal basis. Assuming that $H^{-1}A$ has been chosen (that is, an iteration process has been decided upon), the only free choice remaining to accelerate the removal of the error terms is the choice of h. As we shall see, the three classical methods have all been conditioned by the choice of H to have an optimum h equal to 1 for a stationary iteration process.

9.3.2 ODE Form of the Classical Methods

The three iterative procedures defined by (9.30)–(9.32) obey no apparent pattern except that they are easy to implement in a computer code since all of the data required to update the value of one point are explicitly available at the time of the update. Now let us study these methods as subsets of ODE's as formulated in Section 9.3.1. Insert the model equation (9.29) into the ODE form (9.35). Then

$$H\frac{d\vec{\phi}}{dt} = B(1, -2, 1)\vec{\phi} - \vec{f}. \qquad (9.40)$$

As a start, let us use for the numerical integration the explicit Euler method

$$\phi_{n+1} = \phi_n + h\phi'_n \qquad (9.41)$$

with a step size, h, equal to 1. We arrive at

$$H(\vec{\phi}_{n+1} - \vec{\phi}_n) = B(1, -2, 1)\vec{\phi}_n - \vec{f}. \qquad (9.42)$$

It is clear that the best choice of H from the point of view of matrix algebra is $-B(1, -2, 1)$ since then multiplication from the left by $-B(1, -2, 1)^{-1}$ gives the correct answer in one step. However, this is not in the spirit of our study, since multiplication by the inverse amounts to solving the problem by a direct method without iteration. The constraint on H that is in keeping with the formulation of the three methods described in Section 9.2.3 is that all the elements above the diagonal (or below the diagonal if the sweeps are from right to left) are zero. If we impose this constraint and further restrict ourselves to banded tridiagonals with a single constant in each band, we are led to

$$B(-\beta, \frac{2}{\omega}, 0)(\vec{\phi}_{n+1} - \phi'_n) = B(1, -2, 1)\vec{\phi}_n - \vec{f}, \qquad (9.43)$$

where β and ω are arbitrary. With this choice of notation, the three methods presented in Section 9.2.3 can be identified using the entries in Table 9.1.

Table 9.1. Values of β and ω that lead to the classical relaxation methods

β	ω	Method	Equation
0	1	Point-Jacobi	(9.30)
1	1	Gauss–Seidel	(9.31)
1	$2/\left[1 + \sin\left(\frac{\pi}{M+1}\right)\right]$	Optimum SOR	(9.32)

The fact that the values in the tables lead to the methods indicated can be verified by simple algebraic manipulation. However, our purpose is to examine the whole procedure as a special subset of the theory of ordinary differential equations. In this light, the three methods are all contained in the following set of ODE's

$$\frac{d\vec{\phi}}{dt} = B^{-1}(-\beta, \frac{2}{\omega}, 0)\left[B(1, -2, 1)\vec{\phi} - \vec{f}\right] \tag{9.44}$$

and appear from it in the special case when the explicit Euler method is used for its numerical integration. The point operator that results from the use of the explicit Euler scheme is

$$\phi_j^{(n+1)} = \left(\frac{\omega\beta}{2}\phi_{j-1}^{(n+1)} + \frac{\omega}{2}(h-\beta)\phi_{j-1}^{(n)}\right) - \left((\omega h - 1)\phi_j^{(n)}\right) + \left(\frac{\omega h}{2}\phi_{j+1}^{(n)}\right)$$
$$-\frac{\omega h}{2}f_j . \tag{9.45}$$

This represents a generalization of the classical relaxation techniques.

9.4 Eigensystems of the Classical Methods

The ODE approach to relaxation can be summarized as follows. The basic equation to be solved came from some time-accurate derivation

$$A_b\vec{u} - \vec{f_b} = 0 . \tag{9.46}$$

This equation is preconditioned in some manner which has the effect of multiplication by a conditioning matrix C, giving

$$A\vec{\phi} - \vec{f} = 0 . \tag{9.47}$$

An iterative scheme is developed to find the converged, or steady-state, solution of the set of ODE's

$$H\frac{d\vec{\phi}}{dt} = A\vec{\phi} - \vec{f} . \tag{9.48}$$

This solution has the analytical form

$$\vec{\phi}_n = \vec{e}_n + \vec{\phi}_\infty , \tag{9.49}$$

where $\vec{e}_n$ is the transient, or error, and $\vec{\phi}_\infty \equiv A^{-1}\vec{f}$ is the steady-state solution. The three classical methods, point-Jacobi, Gauss–Seidel, and SOR, are identified for the one-dimensional case by (9.44) (solved using the explicit Euler method with $h = 1$) and Table 9.1.

Given our assumption that the component of the error associated with each eigenvector is equally likely to be excited, the asymptotic convergence rate is determined by the eigenvalue σ_m of $G\ (\equiv I + H^{-1}A)$ having maximum absolute value. Thus

$$\text{Convergence rate} \sim |\sigma_m|_{\max} , \quad m = 1, 2, \cdots, M . \tag{9.50}$$

In this section, we use the ODE analysis to find the convergence rates of the three classical methods represented by (9.30)–(9.32). It is also instructive to inspect the eigenvectors and eigenvalues in the $H^{-1}A$ matrix for the three methods. This amounts to solving the generalized eigenvalue problem

$$A\vec{x}_m = \lambda_m H \vec{x}_m \tag{9.51}$$

for the special case

$$B(1, -2, 1)\vec{x}_m = \lambda_m B\left(-\beta, \frac{2}{\omega}, 0\right)\vec{x}_m . \tag{9.52}$$

The generalized eigensystem for simple tridiagonals is given in Appendix B.2. The three special cases considered below are obtained with $a = 1$, $b = -2$, $c = 1$, $d = -\beta$, $e = 2/\omega$, and $f = 0$. To illustrate the behavior, we take $M = 5$ for the matrix order. This special case makes the general result quite clear.

9.4.1 The Point-Jacobi System

If $\beta = 0$ and $\omega = 1$ in (9.44), the ODE matrix $H^{-1}A$ reduces to simply $B(\frac{1}{2}, -1, \frac{1}{2})$. The eigensystem can be determined from Appendix B.1 since both d and f are zero. The eigenvalues are given by the equation

$$\lambda_m = -1 + \cos\left(\frac{m\pi}{M+1}\right), \quad m = 1, 2, \ldots, M . \tag{9.53}$$

The λ–σ relation for the explicit Euler method is $\sigma_m = 1 + \lambda_m h$. This relation can be plotted for any h. The plot for $h = 1$, the optimum stationary case, is shown in Figure 9.1. For $h < 1$, the maximum $|\sigma_m|$ is obtained with $m = 1$, as shown in Figure 9.2 and for $h > 1$, the maximum $|\sigma_m|$ is obtained with $m = M$, as shown in Figure 9.3. Note that for $h > 1.0$ (depending on M) there is the possibility of instability, i.e. $|\sigma_m| > 1.0$. To obtain the optimal scheme, we wish to minimize the maximum $|\sigma_m|$. This is achieved when $h = 1$, $|\sigma_1| = |\sigma_M|$, and the best possible convergence rate is achieved:

$$|\sigma_m|_{\max} = \cos\left(\frac{\pi}{M+1}\right) . \tag{9.54}$$

For $M = 40$, we obtain $|\sigma_m|_{\max} = 0.9971$. Thus after 500 iterations the error content associated with each eigenvector is reduced to no more than 0.23 times its initial level.

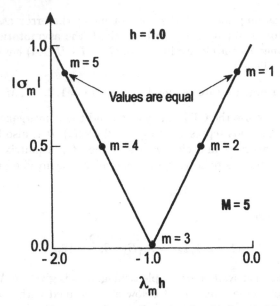

Fig. 9.1. The λ–σ relation for point-Jacobi, $h = 1$, $M = 5$

Again from Appendix B.1, the eigenvectors of $H^{-1}A$ are given by

$$\vec{x}_j = (x_j)_m = \sin\left[j\left(\frac{m\pi}{M+1}\right)\right], \quad j = 1, 2, \ldots, M . \qquad (9.55)$$

This is a very "well-behaved" eigensystem with linearly independent eigenvectors and distinct eigenvalues. The first five eigenvectors are simple sine waves. For $M = 5$, the eigenvectors can be written as

$$\vec{x}_1 = \begin{bmatrix} 1/2 \\ \sqrt{3}/2 \\ 1 \\ \sqrt{3}/2 \\ 1/2 \end{bmatrix}, \quad \vec{x}_2 = \begin{bmatrix} \sqrt{3}/2 \\ \sqrt{3}/2 \\ 0 \\ -\sqrt{3}/2 \\ -\sqrt{3}/2 \end{bmatrix}, \quad \vec{x}_3 = \begin{bmatrix} 1 \\ 0 \\ -1 \\ 0 \\ 1 \end{bmatrix},$$

$$\vec{x}_4 = \begin{bmatrix} \sqrt{3}/2 \\ -\sqrt{3}/2 \\ 0 \\ \sqrt{3}/2 \\ -\sqrt{3}/2 \end{bmatrix}, \quad \vec{x}_5 = \begin{bmatrix} 1/2 \\ -\sqrt{3}/2 \\ 1 \\ -\sqrt{3}/2 \\ 1/2 \end{bmatrix} . \qquad (9.56)$$

The corresponding eigenvalues are, from (9.53),

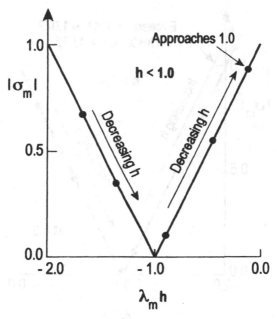

Fig. 9.2. The λ–σ relation for point-Jacobi, $h = 0.9, M = 5$

$$\lambda_1 = -1 + \frac{\sqrt{3}}{2} = -0.134 \cdots$$

$$\lambda_2 = -1 + \tfrac{1}{2} = -0.5$$

$$\lambda_3 = -1 \phantom{+\tfrac{1}{2}} = -1.0 \qquad (9.57)$$

$$\lambda_4 = -1 - \tfrac{1}{2} = -1.5$$

$$\lambda_5 = -1 - \frac{\sqrt{3}}{2} = -1.866 \cdots .$$

From (9.39), the numerical solution written in full is

$$\vec{\phi}_n - \vec{\phi}_\infty = c_1[1 - (1 - \frac{\sqrt{3}}{2})h]^n \vec{x}_1$$
$$+ c_2[1 - (1 - \frac{1}{2})h]^n \vec{x}_2$$
$$+ c_3[1 - h]^n \vec{x}_3$$
$$+ c_4[1 - (1 + \frac{1}{2})h]^n \vec{x}_4$$
$$+ c_5[1 - (1 + \frac{\sqrt{3}}{2})h]^n \vec{x}_5 . \qquad (9.58)$$

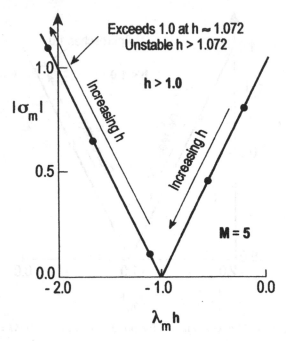

Fig. 9.3. The λ–σ relation for point-Jacobi, $h = 1.1, M = 5$

9.4.2 The Gauss–Seidel System

If β and ω are equal to 1 in (9.44), the matrix eigensystem evolves from the relation

$$B(1, -2, 1)\vec{x}_m = \lambda_m B(-1, 2, 0)\vec{x}_m , \qquad (9.59)$$

which can be studied using the results in Appendix B.2. One can show that the $H^{-1}A$ matrix for the Gauss–Seidel method, A_{GS}, is

$$A_{\mathrm{GS}} \equiv B(-1, 2, 0)^{-1}B(1, -2, 1)$$

$$= \begin{bmatrix} -1 & 1/2 & & & & \\ 0 & -3/4 & 1/2 & & & \\ 0 & 1/8 & -3/4 & 1/2 & & \\ 0 & 1/16 & 1/8 & -3/4 & 1/2 & \\ 0 & 1/32 & 1/16 & 1/8 & -3/4 & 1/2 \\ \vdots & \vdots & \ddots & \ddots & \ddots & \ddots & \ddots \\ 0 & 1/2^M & \cdots & & & \end{bmatrix} \begin{matrix} 1 \\ 2 \\ 3 \\ 4 \\ 5 \\ \vdots \\ M \end{matrix} . \qquad (9.60)$$

The eigenvector structure of the Gauss–Seidel ODE matrix is quite interesting. If M is odd, there are $(M + 1)/2$ distinct eigenvalues with corresponding linearly independent eigenvectors, and there are $(M-1)/2$ defective

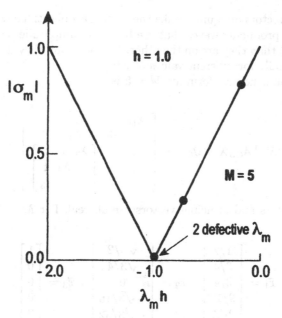

Fig. 9.4. The λ–σ relation for Gauss–Seidel, $h = 1.0, M = 5$

eigenvalues with corresponding principal vectors. The equation for the non-defective eigenvalues in the ODE matrix is (for odd M)

$$\lambda_m = -1 + \cos^2\left(\frac{m\pi}{M+1}\right) \quad , \quad m = 1, 2, \ldots, \frac{M+1}{2} \tag{9.61}$$

and the corresponding eigenvectors are given by

$$\vec{x}_m = \left[\cos\left(\frac{m\pi}{M+1}\right)\right]^{j-1} \sin\left[j\left(\frac{m\pi}{M+1}\right)\right] \quad ,$$
$$m = 1, 2, \ldots, \frac{M+1}{2} . \tag{9.62}$$

The λ–σ relation for $h = 1$, the optimum stationary case, is shown in Figure 9.4. The σ_m with the largest amplitude is obtained with $m = 1$. Hence the convergence rate is

$$|\sigma_m|_{\max} = \left[\cos\left(\frac{\pi}{M+1}\right)\right]^2 . \tag{9.63}$$

Since this is the square of that obtained for the point-Jacobi method, the error associated with the "worst" eigenvector is removed in half as many iterations. For $M = 40$, $|\sigma_m|_{\max} = 0.9942$, and 250 iterations are required to reduce the error component of the worst eigenvector by a factor of roughly 0.23.

The eigenvectors are quite unlike the point-Jacobi set. They are no longer symmetrical, producing waves that are higher in amplitude on one side (the updated side) than they are on the other. Furthermore, they do not represent a common family for different values of M.

The Jordan canonical form for $M = 5$ is

$$X^{-1}A_{GS}X = J_{GS} = \begin{bmatrix} [\lambda_1] & & & \\ & [\lambda_2] & & \\ & & \begin{matrix} \lambda_3 & 1 & \\ & \lambda_3 & 1 \\ & & \lambda_3 \end{matrix} \end{bmatrix}. \tag{9.64}$$

The eigenvectors and principal vectors are all real. For $M = 5$ they can be written

$$\vec{x}_1 = \begin{bmatrix} 1/2 \\ 3/4 \\ 3/4 \\ 9/16 \\ 9/32 \end{bmatrix}, \; \vec{x}_2 = \begin{bmatrix} \sqrt{3}/2 \\ \sqrt{3}/4 \\ 0 \\ -\sqrt{3}/16 \\ -\sqrt{3}/32 \end{bmatrix}, \; \vec{x}_3 = \begin{bmatrix} 1 \\ 0 \\ 0 \\ 0 \\ 0 \end{bmatrix},$$

$$\vec{x}_4 = \begin{bmatrix} 0 \\ 2 \\ -1 \\ 0 \\ 0 \end{bmatrix}, \; \vec{x}_5 = \begin{bmatrix} 0 \\ 0 \\ 4 \\ -4 \\ 1 \end{bmatrix}. \tag{9.65}$$

The corresponding eigenvalues are

$$\begin{aligned} \lambda_1 &= -1/4 \\ \lambda_2 &= -3/4 \\ \lambda_3 &= -1 \\ \left.\begin{matrix}(4) \\ (5)\end{matrix}\right\} & \quad \text{defective, linked to } \lambda_3 \\ & \quad \text{Jordan block.} \end{aligned} \tag{9.66}$$

The numerical solution written in full is thus

$$\begin{aligned} \vec{\phi}_n - \vec{\phi}_\infty = & \; c_1(1 - \frac{h}{4})^n \vec{x}_1 \\ & + c_2(1 - \frac{3h}{4})^n \vec{x}_2 \\ & + \left[c_3(1-h)^n + c_4 h \frac{n}{1!}(1-h)^{n-1} + c_5 h^2 \frac{n(n-1)}{2!}(1-h)^{n-2}\right]\vec{x}_3 \\ & + \left[c_4(1-h)^n + c_5 h \frac{n}{1!}(1-h)^{n-1}\right]\vec{x}_4 \\ & + c_5(1-h)^n \vec{x}_5. \end{aligned} \tag{9.67}$$

9.4.3 The SOR System

If $\beta = 1$ and $2/\omega = x$ in (9.44), the ODE matrix is $B(-1, x, 0)^{-1} B(1, -2, 1)$. One can show that this can be written in the form given below for $M = 5$. The generalization to any M is fairly clear. The $H^{-1} A$ matrix for the SOR method, $A_{\mathrm{SOR}} \equiv B(-1, x, 0)^{-1} B(1, -2, 1)$, is

$$\frac{1}{x^5}
\begin{bmatrix}
-2x^4 & x^4 & 0 & 0 & 0 \\
-2x^3 + x^4 & x^3 - 2x^4 & x^4 & 0 & 0 \\
-2x^2 + x^3 & x^2 - 2x^3 + x^4 & x^3 - 2x^4 & x^4 & 0 \\
-2x + x^2 & x - 2x^2 + x^3 & x^2 - 2x^3 + x^4 & x^3 - 2x^4 & x^4 \\
-2 + x & 1 - 2x + x^2 & x - 2x^2 + x^3 & x^2 - 2x^3 + x^4 & x^3 - 2x^4
\end{bmatrix} .$$

Eigenvalues of the system are given by

$$\lambda_m = -1 + \left(\frac{\omega p_m + z_m}{2} \right)^2 \;, \quad m = 1, 2, \ldots M \;, \tag{9.68}$$

where

$$z_m = [4(1 - \omega) + \omega^2 p_m^2]^{1/2}$$
$$p_m = \cos[m\pi/(M + 1)] \;.$$

If $\omega = 1$, the system is Gauss–Seidel. If $4(1 - \omega) + \omega^2 p_m^2 < 0$, z_m and λ_m are complex. If ω is chosen such that $4(1 - \omega) + \omega^2 p_1^2 = 0$, ω is optimum for the stationary case, and the following conditions hold:

(1) Two eigenvalues are real, equal, and defective.
(2) If M is even, the remaining eigenvalues are complex and occur in conjugate pairs.
(3) If M is odd, one of the remaining eigenvalues is real, and the others are complex occurring in conjugate pairs.

One can easily show that the optimum ω for the stationary case is

$$\omega_{\mathrm{opt}} = \frac{2}{1 + \sin[\pi/(M + 1)]} \tag{9.69}$$

and, for $\omega = \omega_{\mathrm{opt}}$,

$$\lambda_m = \zeta_m^2 - 1$$
$$\vec{x}_m = \zeta_m^{j-1} \sin\left[j \left(\frac{m\pi}{M + 1} \right) \right] , \tag{9.70}$$

where

$$\zeta_m = \frac{\omega_{\mathrm{opt}}}{2} \left[p_m + i\sqrt{p_1^2 - p_m^2} \right] .$$

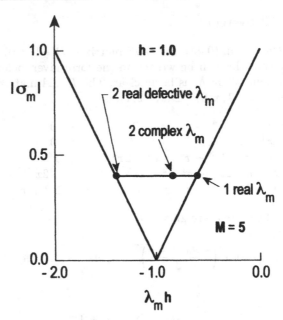

Fig. 9.5. The λ–σ relation for optimum stationary SOR, $M = 5$, $h = 1$

Using the explicit Euler method to integrate the ODE's, $\sigma_m = 1 - h + h\zeta_m^2$, and if $h = 1$, the optimum value for the stationary case, the λ–σ relation reduces to that shown in Figure 9.5. This illustrates the fact that for optimum stationary SOR all the $|\sigma_m|$ are identical and equal to $\omega_{\text{opt}} - 1$. Hence the convergence rate is

$$|\sigma_m|_{\text{max}} = \omega_{\text{opt}} - 1 \tag{9.71}$$

$$\omega_{\text{opt}} = \frac{2}{1 + \sin\left[\pi/(M + 1)\right]} .$$

For $M = 40$, $|\sigma_m|_{\text{max}} = 0.8578$. Hence the worst error component is reduced to less than 0.23 times its initial value in only ten iterations, much faster than both Gauss–Seidel and point-Jacobi. In practical applications, the optimum value of ω may have to be determined by trial and error, and the benefit may not be as great.

For odd M, there are two real eigenvectors and one real principal vector. The remaining linearly independent eigenvectors are all complex. For $M = 5$, they can be written

$$\vec{x}_1 = \begin{bmatrix} 1/2 \\ 1/2 \\ 1/3 \\ 1/6 \\ 1/18 \end{bmatrix} , \quad \vec{x}_2 = \begin{bmatrix} -6 \\ 9 \\ 16 \\ 13 \\ 6 \end{bmatrix} ,$$

$$\vec{x}_{3,4} = \begin{bmatrix} \sqrt{3}(1\quad)/2 \\ \sqrt{3}(1 \pm \text{ i}\sqrt{2})/6 \\ 0 \\ \sqrt{3}(5 \pm \text{ i}\sqrt{2})/54 \\ \sqrt{3}(7 \pm 4\text{i}\sqrt{2})/162 \end{bmatrix}, \quad \vec{x}_5 = \begin{bmatrix} 1 \\ 0 \\ 1/3 \\ 0 \\ 1/9 \end{bmatrix} . \tag{9.72}$$

The corresponding eigenvalues are

$$
\begin{aligned}
\lambda_1 &= -2/3 \\
(2) & \quad \text{defective linked to} \quad \lambda_1 \\
\lambda_3 &= -(10 - 2\sqrt{2}\text{i})/9 \\
\lambda_4 &= -(10 + 2\sqrt{2}\text{i})/9 \\
\lambda_5 &= -4/3 .
\end{aligned}
\tag{9.73}
$$

The numerical solution written in full is

$$
\begin{aligned}
\vec{\phi}_n - \vec{\phi}_\infty = {} & [c_1(1 - 2h/3)^n + c_2 nh(1 - 2h/3)^{n-1}]\vec{x}_1 \\
& + c_2(1 - 2h/3)^n \vec{x}_2 \\
& + c_3[1 - (10 - 2\sqrt{2}\text{i})h/9]^n \vec{x}_3 \\
& + c_4[1 - (10 + 2\sqrt{2}\text{i})h/9]^n \vec{x}_4 \\
& + c_5(1 - 4h/3)^n \vec{x}_5 .
\end{aligned}
\tag{9.74}
$$

9.5 Nonstationary Processes

In classical terminology a method is said to be nonstationary if the conditioning matrices, H and C, are varied at each time step. This does not change the steady-state solution $A_b^{-1} f_b$, but it can greatly affect the convergence rate. In our ODE approach this could also be considered and would lead to a study of equations with nonconstant coefficients. It is much simpler, however, to study the case of fixed H and C but variable step size, h. This process changes the point-Jacobi method to Richardson's method in standard terminology. For the Gauss–Seidel and SOR methods it leads to processes that can be superior to the stationary methods.

The nonstationary form of (9.39) is

$$
\vec{\phi}_N = c_1 \vec{x}_1 \prod_{n=1}^{N}(1 + \lambda_1 h_n) + \cdots + c_m \vec{x}_m \prod_{n=1}^{N}(1 + \lambda_m h_n)
$$

$$
+ \cdots + c_M \vec{x}_M \prod_{n=1}^{N}(1 + \lambda_M h_n) + \vec{\phi}_\infty . \tag{9.75}
$$

where the symbol $\prod$ stands for product. Since h_n can now be changed at each step, the error term can theoretically be completely eliminated in M

steps by taking $h_m = -1/\lambda_m$, for $m = 1, 2, \cdots, M$. However, the eigenvalues λ_m are generally unknown and costly to compute. It is therefore unnecessary and impractical to set $h_m = -1/\lambda_m$ for $m = 1, 2, \ldots, M$. We will see that a few well chosen h's can reduce whole clusters of eigenvectors associated with nearby λ's in the λ_m spectrum. This leads to the concept of selectively annihilating clusters of eigenvectors from the error terms as part of a total iteration process. This is the basis for the multigrid methods discussed in Chapter 10.

Let us consider the very important case when all of the λ_m are *real* and *negative* (remember that they arise from a conditioned matrix so this constraint is not unrealistic for quite practical cases). Consider one of the error terms taken from

$$\vec{e}_N \equiv \vec{\phi}_N - \vec{\phi}_\infty = \sum_{m=1}^{M} c_m \vec{x}_m \prod_{n=1}^{N} (1 + \lambda_m h_n) \qquad (9.76)$$

and write it in the form

$$c_m \vec{x}_m P_e(\lambda_m) \equiv c_m \vec{x}_m \prod_{n=1}^{N} (1 + \lambda_m h_n) , \qquad (9.77)$$

where P_e signifies an "Euler" polynomial. Now focus attention on the polynomial

$$(P_e)_N(\lambda) = (1 + h_1\lambda)(1 + h_2\lambda) \cdots (1 + h_N\lambda) \qquad (9.78)$$

treating it as a continuous function of the independent variable λ. In the annihilation process mentioned after (9.75), we considered making the error exactly zero by taking advantage of some knowledge about the discrete values of λ_m for a particular case. Now we pose a less demanding problem. Let us choose the h_n so that the maximum value of $(P_e)_N(\lambda)$ is as small as possible for all λ lying between λ_a and λ_b such that $\lambda_b \leq \lambda \leq \lambda_a \leq 0$. Mathematically stated, we seek

$$\max_{\lambda_b \leq \lambda \leq \lambda_a} |(P_e)_N(\lambda)| = \text{minimum} \quad , \quad \text{with}(P_e)_N(0) = 1 . \qquad (9.79)$$

This problem has a well known solution due to Markov. It is

$$(P_e)_N(\lambda) = \frac{T_N[(2\lambda - \lambda_a - \lambda_b)/(\lambda_a - \lambda_b)]}{T_N[(-\lambda_a - \lambda_b)/(\lambda_a - \lambda_b)]} , \qquad (9.80)$$

where

$$T_N(y) = \cos(N \arccos y) \qquad (9.81)$$

are the Chebyshev polynomials along the interval $-1 \leq y \leq 1$ and

$$T_N(y) = \frac{1}{2}\left(y + \sqrt{y^2 - 1}\right)^N + \frac{1}{2}\left(y - \sqrt{y^2 - 1}\right)^N \tag{9.82}$$

are the Chebyshev polynomials for $|y| > 1$. In relaxation terminology this is generally referred to as Richardson's method, and it leads to the nonstationary step size choice given by

$$\frac{1}{h_n} = \frac{1}{2}\left\{-\lambda_b - \lambda_a + (\lambda_b - \lambda_a)\cos\left[\frac{(2n-1)\pi}{2N}\right]\right\}, \quad n = 1, 2, \ldots N \ . \tag{9.83}$$

Remember that all λ are negative real numbers representing the magnitudes of λ_m in an eigenvalue spectrum.

The error in the relaxation process represented by (9.75) is expressed in terms of a set of eigenvectors, $\vec{x}_m$, amplified by the coefficients $c_m \prod(1 + \lambda_m h_n)$. With each eigenvector there is a corresponding eigenvalue. Equation (9.83) gives us the best choice of a series of h_n that will minimize the amplitude of the error carried in the eigenvectors associated with the eigenvalues between λ_b and λ_a.

As an example for the use of (9.83), let us consider the following problem:

> Minimize the maximum error associated with the λ eigenvalues in the interval $-2 \le \lambda \le -1$ using only three iterations. $\tag{9.84}$

The three values of h which satisfy this problem are

$$h_n = \frac{2}{3 - \cos\left[(2n-1)\pi/6\right]}, \quad n = 1, 2, 3 \tag{9.85}$$

and the amplitude of the eigenvector is reduced to

$$(P_e)_3(\lambda) = T_3(2\lambda + 3)/T_3(3) \ , \tag{9.86}$$

where

$$T_3(3) = \left[(3 + \sqrt{8})^3 + (3 - \sqrt{8})^3\right]/2 \approx 99 \ . \tag{9.87}$$

A plot of (9.86) is given in Figure 9.6, and we see that the amplitudes of all the eigenvectors associated with the eigenvalues in the range $-2 \le \lambda \le -1$ have been reduced to less than about 1% of their initial values. The values of h used in Figure 9.6 are

$$h_1 = 4/(6 - \sqrt{3})$$
$$h_2 = 4/(6 - 0)$$
$$h_3 = 4/(6 + \sqrt{3}) \ .$$

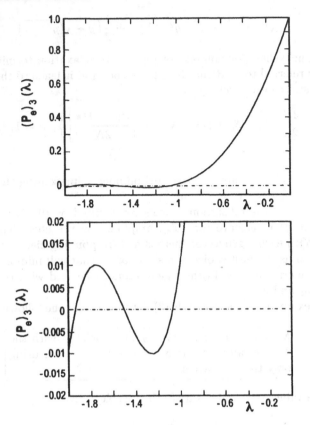

Fig. 9.6. Richardson's method for three steps, minimization over $-2 \leq \lambda \leq -1$

Return now to (9.75). This was derived from (9.37) on the condition that the explicit Euler method, (9.41), was used to integrate the basic ODE's. If instead the implicit trapezoidal rule

$$\phi_{n+1} = \phi_n + \frac{1}{2}h(\phi'_{n+1} + \phi'_n) \tag{9.88}$$

is used, the nonstationary formula

$$\vec{\phi}_N = \sum_{m=1}^{M} c_m \vec{x}_m \prod_{n=1}^{N} \left(\frac{1 + \frac{1}{2}h_n\lambda_m}{1 - \frac{1}{2}h_n\lambda_m} \right) + \vec{\phi}_\infty \tag{9.89}$$

results. This calls for a study of the rational "trapezoidal" polynomial, P_t:

$$(P_t)_N(\lambda) = \prod_{n=1}^{N} \left(\frac{1 + \frac{1}{2}h_n\lambda}{1 - \frac{1}{2}h_n\lambda} \right) \tag{9.90}$$

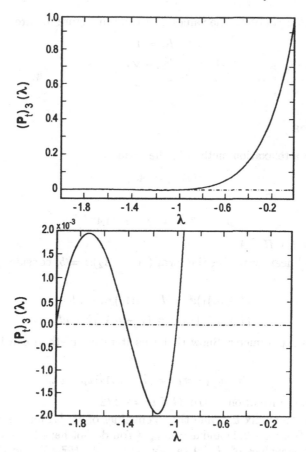

Fig. 9.7. Wachspress method for three steps, minimization over $-2 \leq \lambda \leq -1$

under the same constraints as before, namely that

$$\max_{\lambda_b \leq \lambda \leq \lambda_a} |(P_t)_N(\lambda)| = \text{minimum} \tag{9.91}$$

with

$$(P_t)_N(0) = 1 .$$

The optimum values of h can also be found for this problem, but we settle here for the approximation suggested by Wachspress

$$\frac{2}{h_n} = -\lambda_b \left(\frac{\lambda_a}{\lambda_b} \right)^{(n-1)/(N-1)} , \quad n = 1, 2, \cdots, N . \tag{9.92}$$

This process is also applied to (9.84). The results for $(P_t)_3(\lambda)$ are shown in Figure 9.7. The error amplitude is about 1/5 of that found for $(P_e)_3(\lambda)$ in the same interval of λ. The values of h used in Figure 9.7 are

$$h_1 = 1$$
$$h_2 = \sqrt{2}$$
$$h_3 = 2 \,.$$

Exercises

9.1 Given a relaxation method in the form

$$H \Delta \vec{\phi}_n = A \vec{\phi}_n - \vec{f}$$

show that

$$\vec{\phi}_n = G^n \vec{\phi}_0 + (I - G^n) A^{-1} \vec{f} \,,$$

where $G = I + H^{-1}A$.

9.2 For a linear system of the form $(A_1 + A_2)x = b$, consider the iterative method

$$(I + \mu A_1)\tilde{x} = (I - \mu A_2)x_n + \mu b$$
$$(I + \mu A_2)x_{n+1} = (I - \mu A_1)\tilde{x} + \mu b \,,$$

where μ is a parameter. Show that this iterative method can be written in the form

$$H(x_{k+1} - x_k) = (A_1 + A_2)x_k - b \,.$$

Determine the iteration matrix G if $\mu = -1/2$.

9.3 Using Appendix B.2, find the eigenvalues of $H^{-1}A$ for the SOR method with $A = B(4 : 1, -2, 1)$ and $\omega = \omega_{\text{opt}}$. (You do not have to find H^{-1}. Recall that the eigenvalues of $H^{-1}A$ satisfy $A\tilde{x}_m = \lambda_m H\tilde{x}_m$.) Find the numerical values, not just the expressions. Then find the corresponding $|\sigma_m|$ values.

9.4 Solve the following equation on the domain $0 \le x \le 1$ with boundary conditions $u(0) = 0$, $u(1) = 1$:

$$\frac{\partial^2 u}{\partial x^2} - 6x = 0$$

For the initial condition, use $u(x) = 0$. Use second-order centered differences on a grid with 40 cells ($M = 39$). Iterate to steady state using

(a) the point-Jacobi method,
(b) the Gauss–Seidel method,
(c) the SOR method with the optimum value of ω, and
(d) the 3-step Richardson method derived in Section 9.5.

Plot the solution after the residual is reduced by 2, 3, and 4 orders of magnitude. Plot the logarithm of the L_2-norm of the residual vs. the number of iterations. Determine the asymptotic convergence rate. Compare with the theoretical asymptotic convergence rate.

10. Multigrid

The idea of systematically using sets of coarser grids to accelerate the convergence of iterative schemes that arise from the numerical solution of partial differential equations was made popular by the work of Brandt. There are many variations of the process and many viewpoints of the underlying theory. The viewpoint presented here is a natural extension of the concepts discussed in Chapter 9.

10.1 Motivation

10.1.1 Eigenvector and Eigenvalue Identification with Space Frequencies

Consider the eigensystem of the model matrix $B(1, -2, 1)$. The eigenvalues and eigenvectors are given in Sections 4.3.2 and 4.3.3, respectively. Notice that as the magnitudes of the eigenvalues increase, the space-frequencies (or wavenumbers) of the corresponding eigenvectors also increase. That is, if the eigenvalues are ordered such that

$$|\lambda_1| \leq |\lambda_2| \leq \cdots \leq |\lambda_M| \tag{10.1}$$

then the corresponding eigenvectors are ordered from low to high space-frequencies. This has a rational explanation from the origin of the banded matrix. Note that

$$\frac{\partial^2}{\partial x^2} \sin(mx) = -m^2 \sin(mx) \tag{10.2}$$

and recall that

$$\delta_{xx}\vec{\phi} = \frac{1}{\Delta x^2} B(1, -2, 1)\vec{\phi} = X \left[\frac{1}{\Delta x^2} D(\vec{\lambda}) \right] X^{-1}\vec{\phi}, \tag{10.3}$$

where $D(\vec{\lambda})$ is a diagonal matrix containing the eigenvalues. We have seen that $X^{-1}\vec{\phi}$ represents a sine transform, and $X\vec{\phi}$, a sine synthesis. Therefore, the

operation $\frac{1}{\Delta x^2} D(\vec{\lambda})$ represents the numerical approximation of the multiplication of the appropriate sine wave by the negative square of its wavenumber,
$-m^2$. One finds that

$$\frac{1}{\Delta x^2} \lambda_m = \left(\frac{M+1}{\pi} \right)^2 \left[-2 + 2\cos \left(\frac{m\pi}{M+1} \right) \right] \approx -m^2, \quad m \ll M .$$

(10.4)

Hence, the correlation of large magnitudes of λ_m with high space-frequencies
is to be expected for these particular matrix operators. This is consistent with
the physics of diffusion as well. However, this correlation is not necessary in
general. In fact, the complete counterexample of the above association is
contained in the eigensystem for $B(\frac{1}{2}, 1, \frac{1}{2})$. For this matrix, one finds, from
Appendix B, exactly the opposite behavior.

10.1.2 Properties of the Iterative Method

The second key motivation for multigrid is the following:

- Many iterative methods reduce error components corresponding to eigenvalues of large amplitude more effectively than those corresponding to
  eigenvalues of small amplitude.

This is to be expected of an iterative method which is time accurate. It
is also true, for example, of the Gauss–Seidel method and, by design, of
the Richardson method described in Section 9.5. The classical point-Jacobi
method does not share this property. As we saw in Section 9.4.1, this method
produces the same value of $|\sigma|$ for $\lambda_{\min}$ and $\lambda_{\max}$. However, the property can
be restored by using $h < 1$, as shown in Figure 9.2.

When an iterative method with this property is applied to a matrix with
the above correlation between the modulus of the eigenvalues and the space-
frequency of the eigenvectors, error components corresponding to high space-
frequencies will be reduced more quickly than those corresponding to low
space-frequencies. This is the key concept underlying the multigrid process.

10.2 The Basic Process

First of all we assume that the difference equations representing the basic
partial differential equations are in a form that can be related to a matrix
which has certain basic properties. This form can be arrived at "naturally"
by simply replacing the derivatives in the PDE with difference schemes, as in
the example given by (3.27), or it can be "contrived" by further conditioning,
as in the examples given by (9.11). The basic assumptions required for our
description of the multigrid process are:

(1) The problem is linear.
(2) The eigenvalues, λ_m, of the matrix are all real and negative.
(3) The λ_m are fairly evenly distributed between their maximum and minimum values.
(4) The eigenvectors associated with the eigenvalues having largest magnitudes can be correlated with high frequencies on the differencing mesh.
(5) The iterative procedure used greatly reduces the amplitudes of the eigenvectors associated with eigenvalues in the range between $\frac{1}{2}|\lambda|_{\max}$ and $|\lambda|_{\max}$.

These conditions are sufficient to ensure the validity of the process described next.

Having preconditioned (if necessary) the basic finite-differencing scheme by a procedure equivalent to the multiplication by a matrix C, we are led to the starting formulation

$$C[A_b\vec{\phi}_\infty - \vec{f_b}] = 0 , \qquad (10.5)$$

where the matrix formed by the product CA_b has the properties given above. In (10.5), the vector $\vec{f_b}$ represents the boundary conditions and the forcing function, if any, and $\vec{\phi}_\infty$ is a vector representing the desired exact solution. We start with some initial guess for $\vec{\phi}_\infty$ and proceed through n iterations making use of some iterative process that satisfies property (5) above. We do not attempt to develop an optimum procedure here, but for clarity we suppose that the three-step Richardson method illustrated in Figure 9.7 is used. At the end of the three steps, we find $\vec{r}$, the residual, where

$$\vec{r} = C[A_b\vec{\phi} - \vec{f_b}] . \qquad (10.6)$$

Recall that the $\vec{\phi}$ used to compute $\vec{r}$ is composed of the exact solution $\vec{\phi}_\infty$ and the error $\vec{e}$ in such a way that

$$A\vec{e} - \vec{r} = 0 , \qquad (10.7)$$

where

$$A \equiv CA_b . \qquad (10.8)$$

If one could solve (10.7) for $\vec{e}$ then

$$\vec{\phi}_\infty = \vec{\phi} - \vec{e} . \qquad (10.9)$$

Thus our goal now is to solve for $\vec{e}$. We can write the exact solution for $\vec{e}$ in terms of the eigenvectors of A, and the σ eigenvalues of the Richardson process in the form:

$$\vec{e} = \sum_{m=1}^{M/2} c_m \vec{x}_m \prod_{n=1}^{3} [\sigma(\lambda_m h_n)] + \underbrace{\sum_{m=M/2+1}^{M} c_m \vec{x}_m \prod_{n=1}^{3} [\sigma(\lambda_m h_n)]}_{\text{very low amplitude}} . \quad (10.10)$$

Combining our basic assumptions, we can be sure that the high frequency content of $\vec{e}$ has been greatly reduced (about 1% or less of its original value in the initial guess). In addition, assumption (4) ensures that the error has been *smoothed*.

Next we construct a permutation matrix which separates a vector into two parts, one containing the odd entries, and the other the even entries of the original vector (or any other appropriate sorting which is consistent with the interpolation approximation to be discussed below). For a 7-point example,

$$
\begin{bmatrix} e_2 \\ e_4 \\ e_6 \\ e_1 \\ e_3 \\ e_5 \\ e_7 \end{bmatrix} =
\begin{bmatrix} 0 & 1 & 0 & 0 & 0 & 0 & 0 \\ 0 & 0 & 0 & 1 & 0 & 0 & 0 \\ 0 & 0 & 0 & 0 & 0 & 1 & 0 \\ 1 & 0 & 0 & 0 & 0 & 0 & 0 \\ 0 & 0 & 1 & 0 & 0 & 0 & 0 \\ 0 & 0 & 0 & 0 & 1 & 0 & 0 \\ 0 & 0 & 0 & 0 & 0 & 0 & 1 \end{bmatrix}
\begin{bmatrix} e_1 \\ e_2 \\ e_3 \\ e_4 \\ e_5 \\ e_6 \\ e_7 \end{bmatrix} , \quad
\begin{bmatrix} \vec{e}_e \\ \vec{e}_o \end{bmatrix} = P\vec{e} . \qquad (10.11)
$$

Multiply (10.7) from the left by P and, since a permutation matrix has an inverse which is its transpose, we can write

$$
PA[P^{-1}P]\vec{e} = P\vec{r} . \qquad (10.12)
$$

The operation PAP^{-1} partitions the A matrix to form

$$
\begin{bmatrix} A_1 & A_2 \\ A_3 & A_4 \end{bmatrix}
\begin{bmatrix} \vec{e}_e \\ \vec{e}_o \end{bmatrix} =
\begin{bmatrix} \vec{r}_e \\ \vec{r}_o \end{bmatrix} . \qquad (10.13)
$$

Notice that

$$
A_1\vec{e}_e + A_2\vec{e}_o = \vec{r}_e \qquad (10.14)
$$

is an exact expression.

At this point we make our one crucial assumption. It is that there is some connection between $\vec{e}_e$ and $\vec{e}_o$ brought about by the smoothing property of the Richardson relaxation procedure. Since the top half of the frequency spectrum has been removed, it is reasonable to suppose that the odd points are the average of the even points. For example,

$$
e_1 \approx \frac{1}{2}(e_a + e_2)
$$
$$
e_3 \approx \frac{1}{2}(e_2 + e_4)
$$
$$
e_5 \approx \frac{1}{2}(e_4 + e_6) \qquad \text{or} \quad \vec{e}_o = A_2'\vec{e}_e \qquad (10.15)
$$
$$
e_7 \approx \frac{1}{2}(e_6 + e_b) .
$$

It is important to notice that e_a and e_b represent errors on the boundaries where the error is zero if the boundary conditions are given. It is also important to notice that we are dealing with the relation between $\vec{e}$ and $\vec{r}$, so the original boundary conditions and forcing function (which are contained in $\vec{f}$ in the basic formulation) no longer appear in the problem. Hence, no aliasing of these functions can occur in subsequent steps. Finally, notice that, in this formulation, the averaging of $\vec{e}$ is our only approximation; no operations on $\vec{r}$ are required or justified.

If the boundary conditions are Dirichlet, e_a and e_b are zero, and one can write for the example case

$$A'_2 = \frac{1}{2} \begin{bmatrix} 1 & 0 & 0 \\ 1 & 1 & 0 \\ 0 & 1 & 1 \\ 0 & 0 & 1 \end{bmatrix} . \tag{10.16}$$

With this approximation (10.14) reduces to

$$A_1 \vec{e}_e + A_2 A'_2 \vec{e}_e = \vec{r}_e \tag{10.17}$$

or

$$A_c \vec{e}_e - \vec{r}_e = 0 , \tag{10.18}$$

where

$$A_c = [A_1 + A_2 A'_2] . \tag{10.19}$$

The form of A_c, the matrix on the coarse mesh, is completely determined by the choice of the permutation matrix and the interpolation approximation. If the matrix A is $B(7:1,-2,1)$, our 7-point example produces

$$PAP^{-1} = \begin{bmatrix} -2 & & & & 1 & 1 & \\ & -2 & & & & 1 & 1 \\ & & -2 & & & & 1 & 1 \\ 1 & & & -2 & & & \\ 1 & 1 & & & -2 & & \\ & 1 & 1 & & & -2 & \\ & & 1 & & & & -2 \end{bmatrix} = \begin{bmatrix} A_1 & A_2 \\ A_3 & A_4 \end{bmatrix} \tag{10.20}$$

and (10.18) gives

$$\overbrace{\begin{bmatrix} -2 & & \\ & -2 & \\ & & -2 \end{bmatrix}}^{A_1} + \overbrace{\begin{bmatrix} 1 & 1 & \\ & 1 & 1 \\ & & 1 & 1 \end{bmatrix}}^{A_2} \cdot \overbrace{\frac{1}{2}\begin{bmatrix} 1 & \\ 1 & 1 \\ & 1 & 1 \\ & & 1 \end{bmatrix}}^{A'_2} = \overbrace{\begin{bmatrix} -1 & 1/2 & \\ 1/2 & -1 & 1/2 \\ & 1/2 & -1 \end{bmatrix}}^{A_c} . \tag{10.21}$$

If the boundary conditions are mixed Dirichlet–Neumann, A in the 1D model equation is $B(1, \vec{b}, 1)$ where $\vec{b} = [-2, -2, ..., -2, -1]^{\mathrm{T}}$. The eigensystem is given by (B.18). It is easy to show that the high space-frequencies still correspond to the eigenvalues with high magnitudes, and, in fact, all of the properties given in Section 10.1 are met. However, the eigenvector structure is different from that given in (9.55) for Dirichlet conditions. In the present case they are given by

$$x_{jm} = \sin\left[j\left(\frac{(2m-1)\pi}{2M+1}\right)\right] \quad , \quad m = 1, 2, \cdots, M \; . \tag{10.22}$$

All of them go through zero on the left (Dirichlet) side, and all of them reflect on the right (Neumann) side.

For Neumann conditions, the interpolation formula in (10.15) must be changed. In the particular case given by (10.22), e_b is equal to e_M. If Neumann conditions are on the left, $e_a = e_1$. When $e_b = e_M$, the example in (10.16) changes to

$$A_2' = \frac{1}{2}\begin{bmatrix} 1 & 0 & 0 \\ 1 & 1 & 0 \\ 0 & 1 & 1 \\ 0 & 0 & 2 \end{bmatrix} . \tag{10.23}$$

The permutation matrix remains the same, and both A_1 and A_2 in the partitioned matrix PAP^{-1} are unchanged (only A_4 is modified by putting -1 in the lower right element). Therefore, we can construct the coarse matrix from

$$\overbrace{\begin{bmatrix} -2 & & \\ & -2 & \\ & & -2 \end{bmatrix}}^{A_1} + \overbrace{\begin{bmatrix} 1 & 1 & \\ & 1 & 1 \\ & & 1 & 1 \end{bmatrix}}^{A_2} \cdot \frac{1}{2}\overbrace{\begin{bmatrix} 1 & & \\ 1 & 1 & \\ & 1 & 1 \\ & & 2 \end{bmatrix}}^{A_2'} = \overbrace{\begin{bmatrix} -1 & 1/2 & \\ 1/2 & -1 & 1/2 \\ & 1/2 & -1/2 \end{bmatrix}}^{A_c} ,$$

$$\tag{10.24}$$

which gives us what we might have "expected".

We will continue with Dirichlet boundary conditions for the remainder of this section. At this stage, we have reduced the problem from $B(1, -2, 1)\vec{e} = \vec{r}$ on the fine mesh to $\frac{1}{2}B(1, -2, 1)\vec{e}_e = \vec{r}_e$ on the next coarser mesh. Recall that our goal is to solve for $\vec{e}$, which will provide us with the solution $\vec{\phi}_\infty$ using (10.9). Given $\vec{e}_e$ computed on the coarse grid (possibly using even coarser grids), we can compute $\vec{e}_o$ using (10.15), and thus $\vec{e}$. In order to complete the process, we must now determine the relationship between $\vec{e}_e$ and $\vec{e}$.

In order to examine this relationship, we need to consider the eigensystems of A and A_c:

$$A = X\Lambda X^{-1}, \qquad A_c = X_c \Lambda_c X_c^{-1} . \qquad (10.25)$$

For $A = B(M : 1, -2, 1)$, the eigenvalues and eigenvectors are

$$\lambda_m = -2 \left[1 - \cos\left(\frac{m\pi}{M+1} \right) \right], \quad \vec{x}_m = \sin\left[j \left(\frac{m\pi}{M+1} \right) \right]$$
$$j = 1, 2, \cdots, M \qquad (10.26)$$
$$m = 1, 2, \cdots, M .$$

Based on our assumptions, the most difficult error mode to eliminate is that with $m = 1$, corresponding to

$$\lambda_1 = -2 \left[1 - \cos\left(\frac{\pi}{M+1} \right) \right], \quad \vec{x}_1 = \sin\left[j \left(\frac{\pi}{M+1} \right) \right],$$
$$j = 1, 2, \cdots, M . \qquad (10.27)$$

For example, with $M = 51$, $\lambda_1 = -0.003649$. If we restrict our attention to odd M, then $M_c = (M-1)/2$ is the size of A_c. The eigenvalue and eigenvector corresponding to $m = 1$ for the matrix $A_c = \frac{1}{2} B(M_c, 1, -2, 1)$ are

$$(\lambda_c)_1 = - \left[1 - \cos\left(\frac{2\pi}{M+1} \right) \right], \quad (\vec{x}_c)_1 = \sin\left[j \left(\frac{2\pi}{M+1} \right) \right],$$
$$j = 1, 2, \cdots, M_c . \qquad (10.28)$$

For $M = 51$ ($M_c = 25$), we obtain $(\lambda_c)_1 = -0.007291 = 1.998\lambda_1$. As M increases, $(\lambda_c)_1$ approaches $2\lambda_1$. In addition, one can easily see that $(\vec{x}_c)_1$ coincides with $\vec{x}_1$ at every second point of the latter vector, that is, it contains the even elements of $\vec{x}_1$.

Now let us consider the case in which all of the error consists of the eigenvector component $\vec{x}_1$, i.e. $\vec{e} - \vec{x}_1$. Then the residual is

$$\vec{r} = A\vec{x}_1 = \lambda_1 \vec{x}_1 \qquad (10.29)$$

and the residual on the coarse grid is

$$\vec{r}_e = \lambda_1 (\vec{x}_c)_1 \qquad (10.30)$$

since $(\vec{x}_c)_1$ contains the even elements of $\vec{x}_1$. The exact solution on the coarse grid satisfies

$$\vec{e}_e = A_c^{-1}\vec{r}_e = X_c \Lambda_c^{-1} X_c^{-1} \lambda_1 (\vec{x}_c)_1 \qquad (10.31)$$

$$= \lambda_1 X_c \Lambda_c^{-1} \begin{bmatrix} 1 \\ 0 \\ \vdots \\ 0 \end{bmatrix} \qquad (10.32)$$

$$= \lambda_1 X_c \begin{bmatrix} 1/(\lambda_c)_1 \\ 0 \\ \vdots \\ 0 \end{bmatrix}$$

(10.33)

$$= \frac{\lambda_1}{(\lambda_c)_1} (\vec{x}_c)_1$$

(10.34)

$$\approx \frac{1}{2} (\vec{x}_c)_1 \ .$$

(10.35)

Since our goal is to compute $\vec{e} = \vec{x}_1$, in addition to interpolating $\vec{e}_e$ to the fine grid (using (10.15)), we must *multiply the result by 2*. This is equivalent to solving

$$\frac{1}{2} A_c \vec{e}_e = \vec{r}_e$$

(10.36)

or

$$\frac{1}{4} B(M_c : 1, -2, 1)\vec{e}_e = \vec{r}_e \ .$$

(10.37)

In our case, the matrix $A = B(M : 1, -2, 1)$ comes from a discretization of the diffusion equation, which gives

$$A_b = \frac{\nu}{\Delta x^2} B(M : 1, -2, 1)$$

(10.38)

and the preconditioning matrix C is simply

$$C = \frac{\Delta x^2}{\nu} I \ .$$

(10.39)

Applying the discretization on the coarse grid with the same preconditioning matrix as used on the fine grid gives, since $\Delta x_c = 2\Delta x$,

$$C \frac{\nu}{\Delta x_c^2} B(M_c : 1, -2, 1) = \frac{\Delta x^2}{\Delta x_c^2} B(M_c : 1, -2, 1)$$

$$= \frac{1}{4} B(M_c : 1, -2, 1) \ ,$$

(10.40)

which is precisely the matrix appearing in (10.37). Thus we see that the process is recursive. The problem to be solved on the coarse grid is *the same as that solved on the fine grid.*

The remaining steps required to complete an entire multigrid process are relatively straightforward, but they vary depending on the problem and the user. The reduction can be, and usually is, carried to even coarser grids before returning to the finest level. However, in each case, the appropriate permutation matrix and the interpolation approximation define both the down- and up-going paths. The details of finding optimum techniques are, obviously, quite important but they are not discussed here.

10.3 A Two-Grid Process

We now describe a two-grid process for the linear problem $A\vec{\phi} = \vec{f}$, which can be easily generalized to a process with an arbitrary number of grids due to the recursive nature of multigrid. Extension to nonlinear problems requires that both the solution and the residual be transferred to the coarse grid in a process known as *full approximation storage* multigrid.

(1) Perform n_1 iterations of the selected relaxation method on the fine grid, starting with $\vec{\phi} = \vec{\phi}_n$. Call the result $\vec{\phi}^{(1)}$. This gives[1]

$$\vec{\phi}^{(1)} = G_1^{n_1} \vec{\phi}_n + (I - G_1^{n_1}) A^{-1} \vec{f}, \qquad (10.41)$$

where

$$G_1 = I + H^{-1} A \qquad (10.42)$$

and H is defined as in Chapter 9 (e.g. (9.21)). Next compute the residual based on $\vec{\phi}^{(1)}$:

$$\begin{aligned} r^{(1)} = A\vec{\phi}^{(1)} - \vec{f} &= AG_1^{n_1} \vec{\phi}_n + A(I - G_1^{n_1}) A^{-1} \vec{f} - \vec{f} \\ &= AG_1^{n_1} \vec{\phi}_n - AG_1^{n_1} A^{-1} \vec{f}. \end{aligned} \qquad (10.43)$$

(2) Transfer (or *restrict*) $r^{(1)}$ to the coarse grid:

$$\vec{r}^{(2)} = R_{1\to 2} \vec{r}^{(1)}. \qquad (10.44)$$

For our example in the preceding section, the restriction matrix is

$$R_{1\to 2} = \begin{bmatrix} 0 & 1 & 0 & 0 & 0 & 0 & 0 \\ 0 & 0 & 0 & 1 & 0 & 0 & 0 \\ 0 & 0 & 0 & 0 & 0 & 1 & 0 \end{bmatrix}, \qquad (10.45)$$

that is, the first three rows of the permutation matrix P in (10.11). This type of restriction is known as "simple injection". Some form of weighted restriction can also be used.

---

[1] See Exercise 9.1.

(3) Solve the problem $A_2 \vec{e}^{(2)} = \vec{r}^{(2)}$ on the coarse grid *exactly* (in practice, using Gaussian elimination):[2]

$$\vec{e}^{(2)} = A_2^{-1} \vec{r}^{(2)} . \tag{10.46}$$

Here A_2 can be formed by applying the discretization on the coarse grid. In the preceding example ((10.40)), $A_2 = \frac{1}{4} B(M_c : 1, -2, 1)$. It is at this stage that the generalization to a multigrid procedure with more than two grids occurs. If this is the coarsest grid in the sequence, solve exactly. Otherwise, apply the two-grid process recursively.

(4) Transfer (or *prolong*) the error back to the fine grid and update the solution:

$$\vec{\phi}_{n+1} = \vec{\phi}^{(1)} - I_{2 \to 1} \vec{e}^{(2)} . \tag{10.47}$$

For our example, the prolongation matrix is

$$I_{2 \to 1} = \begin{bmatrix} 1/2 & 0 & 0 \\ 1 & 0 & 0 \\ 1/2 & 1/2 & 0 \\ 0 & 1 & 0 \\ 0 & 1/2 & 1/2 \\ 0 & 0 & 1 \\ 0 & 0 & 1/2 \end{bmatrix} , \tag{10.48}$$

which follows from (10.15).

Combining these steps, one obtains

$$\vec{\phi}_{n+1} = [I - I_{2 \to 1} A_2^{-1} R_{1 \to 2} A] G_1^{n_1} \vec{\phi}_n$$
$$- [I - I_{2 \to 1} A_2^{-1} R_{1 \to 2} A] G_1^{n_1} A^{-1} \vec{f} + A^{-1} \vec{f} . \tag{10.49}$$

Thus the basic iteration matrix is

$$[I - I_{2 \to 1} A_2^{-1} R_{1 \to 2} A] G_1^{n_1} . \tag{10.50}$$

The eigenvalues of this matrix determine the convergence rate of the two-grid process.

The basic iteration matrix for a three-grid process is found from (10.50) by replacing A_2^{-1} with $(I - G_{32}) A_2^{-1}$, where

$$G_{32} = [I - I_{3 \to 2} A_3^{-1} R_{2 \to 3} A_2] G_2^{n_2} . \tag{10.51}$$

In this expression, G_2 is the iteration matrix for the relaxation method on grid 2, analogous to G_1, n_2 is the number of relaxation steps on grid 2, $I_{3 \to 2}$ and $R_{2 \to 3}$ are the transfer operators between grids 2 and 3, and A_3 is obtained by discretizing on grid 3. Extension to four or more grids proceeds in similar fashion.

---

[2] Note that the coarse grid matrix denoted A_2 here was denoted A_c in the preceding section.

Exercises

10.1 Derive (10.51).

10.2 Repeat Exercise 9.4 using a four-grid multigrid method together with

(a) the Gauss–Seidel method,
(b) the 3-step Richardson method derived in Section 9.5.

Solve exactly on the coarsest grid. Plot the solution after the residual is reduced by 2, 3, and 4 orders of magnitude. Plot the logarithm of the L_2-norm of the residual vs. the number of iterations. Determine the asymptotic convergence rate. Calculate the theoretical asymptotic convergence rate and compare.

11. Numerical Dissipation

Up to this point, we have emphasized the second-order centered-difference approximations to the spatial derivatives in our model equations. We have seen that a centered approximation to a first derivative is nondissipative, i.e. the eigenvalues of the associated circulant matrix (with periodic boundary conditions) are pure imaginary. In processes governed by nonlinear equations, such as the Euler and Navier–Stokes equations, there can be a continual production of high-frequency components of the solution, leading, for example, to the formation of shock waves. In a real physical problem, the production of high frequencies is eventually limited by viscosity. However, when we solve the Euler equations numerically, we have neglected viscous effects. Thus the numerical approximation must contain some inherent dissipation to limit the production of high-frequency modes. Although numerical approximations to the Navier–Stokes equations contain dissipation through the viscous terms, this can be insufficient, especially at high Reynolds numbers, due to the limited grid resolution which is practical. Therefore, unless the relevant length scales are resolved, some form of added numerical dissipation is required in the numerical solution of the Navier–Stokes equations as well. Since the addition of numerical dissipation is tantamount to intentionally introducing nonphysical behavior, it must be carefully controlled such that the error introduced is not excessive. In this chapter, we discuss some different ways of adding numerical dissipation to the spatial derivatives in the linear convection equation and hyperbolic systems of PDE's.

11.1 One-Sided First-Derivative Space Differencing

We investigate the properties of one-sided spatial difference operators in the context of the biconvection model equation given by

$$\frac{\partial u}{\partial t} = -a\frac{\partial u}{\partial x} \tag{11.1}$$

with periodic boundary conditions. Consider the following point operator for the spatial derivative term

$$-a(\delta_x u)_j = \frac{-a}{2\Delta x}[-(1+\beta)u_{j-1} + 2\beta u_j + (1-\beta)u_{j+1}]$$

$$= \frac{-a}{2\Delta x}[(-u_{j-1}+u_{j+1}) + \beta(-u_{j-1}+2u_j-u_{j+1})]. \quad (11.2)$$

The second form shown divides the operator into an antisymmetric component $(-u_{j-1}+u_{j+1})/2\Delta x$ and a symmetric component $\beta(-u_{j-1}+2u_j-u_{j+1})/2\Delta x$. The antisymmetric component is the second-order centered difference operator. With $\beta \neq 0$, the operator is only first-order accurate. A *backward* difference operator is given by $\beta = 1$, and a *forward* difference operator is given by $\beta = -1$.

For periodic boundary conditions, the corresponding matrix operator is

$$-a\delta_x = \frac{-a}{2\Delta x}B_p(-1-\beta, 2\beta, 1-\beta).$$

The eigenvalues of this matrix are

$$\lambda_m = \frac{-a}{\Delta x}\left\{\beta\left[1-\cos\left(\frac{2\pi m}{M}\right)\right] + i\sin\left(\frac{2\pi m}{M}\right)\right\}, \quad m = 0, 1, \ldots, M-1.$$

If a is positive, the forward difference operator $(\beta = -1)$ produces $\Re(\lambda_m) > 0$, the centered difference operator $(\beta = 0)$ produces $\Re(\lambda_m) = 0$, and the backward difference operator produces $\Re(\lambda_m) < 0$. Hence the forward difference operator is inherently unstable, while the centered and backward operators are inherently stable. If a is negative, the roles are reversed. When $\Re(\lambda_m) \neq 0$, the solution will either grow or decay with time. In either case, our choice of differencing scheme produces nonphysical behavior. We proceed next to show why this occurs.

11.2 The Modified Partial Differential Equation

First carry out a Taylor series expansion of the terms in (11.2). We are led to the expression

$$(\delta_x u)_j = \frac{1}{2\Delta x}\left[2\Delta x\left(\frac{\partial u}{\partial x}\right)_j - \beta\Delta x^2\left(\frac{\partial^2 u}{\partial x^2}\right)_j + \frac{\Delta x^3}{3}\left(\frac{\partial^3 u}{\partial x^3}\right)_j\right.$$
$$\left. - \frac{\beta\Delta x^4}{12}\left(\frac{\partial^4 u}{\partial x^4}\right)_j + \cdots\right]. \quad (11.3)$$

We see that the antisymmetric portion of the operator introduces odd derivative terms in the truncation error, while the symmetric portion introduces even derivatives. Substituting this into (11.1) gives

$$\frac{\partial u}{\partial t} = -a\frac{\partial u}{\partial x} + \frac{a\beta\Delta x}{2}\frac{\partial^2 u}{\partial x^2} - \frac{a\Delta x^2}{6}\frac{\partial^3 u}{\partial x^3} + \frac{a\beta\Delta x^3}{24}\frac{\partial^4 u}{\partial x^4} + \cdots. \quad (11.4)$$

This is the partial differential equation we are really solving when we apply the approximation given by (11.2) to (11.1). Notice that (11.4) is *consistent* with (11.1), since the two equations are identical when $\Delta x \to 0$. However, when we use a computer to find a numerical solution of the problem, Δx can be small, but it is *not* zero. This means that each term in the expansion given by (11.4) is excited to some degree. We refer to (11.4) as the *modified partial differential equation*. We proceed next to investigate the implications of this concept.

Consider the simple linear partial differential equation

$$\frac{\partial u}{\partial t} = -a\frac{\partial u}{\partial x} + \nu\frac{\partial^2 u}{\partial x^2} + \gamma\frac{\partial^3 u}{\partial x^3} + \tau\frac{\partial^4 u}{\partial x^4} . \tag{11.5}$$

Choose periodic boundary conditions and impose an initial condition $u = e^{i\kappa x}$. Under these conditions, there is a wave-like solution to (11.5) of the form

$$u(x,t) = e^{i\kappa x}e^{(r+is)t}$$

provided r and s satisfy the condition

$$r + is = -ia\kappa - \nu\kappa^2 - i\gamma\kappa^3 + \tau\kappa^4$$

or

$$r = -\kappa^2(\nu - \tau\kappa^2), \quad s = -\kappa(a + \gamma\kappa^2) .$$

The solution is composed of both amplitude and phase terms. Thus

$$u = \underbrace{e^{-\kappa^2(\nu-\tau\kappa^2)t}}_{\text{amplitude}} \; \underbrace{e^{i\kappa[x-(a+\gamma\kappa^2)t]}}_{\text{phase}} . \tag{11.6}$$

It is important to notice that the amplitude of the solution depends only upon ν and τ, the coefficients of the even derivatives in (11.5), and the phase depends only on a and γ, the coefficients of the odd derivatives.

If the wave speed a is positive, the choice of a backward difference scheme ($\beta = 1$) produces a modified PDE with $\nu - \tau\kappa^2 > 0$ and hence the amplitude of the solution decays. This is tantamount to *deliberately adding* dissipation to the PDE. Under the same condition, the choice of a forward difference scheme ($\beta = -1$) is equivalent to *deliberately adding* a destabilizing term to the PDE.

By examining the term governing the phase of the solution in (11.6), we see that the speed of propagation is $a + \gamma\kappa^2$. Referring to the modified PDE, (11.4), we have $\gamma = -a\Delta x^2/6$. Therefore, the phase speed of the numerical solution is *less* than the actual phase speed. Furthermore, the numerical phase speed is dependent upon the wavenumber κ. This we refer to as *dispersion*.

Our purpose here is to investigate the properties of one-sided spatial differencing operators relative to centered difference operators. We have seen that the three-point centered difference approximation of the spatial derivative *produces a modified PDE that has no dissipation* (or amplification). One can easily show, by using the antisymmetry of the matrix difference operators, that the same is true for *any* centered difference approximation of a first derivative. As a corollary, *any departure from antisymmetry in the matrix difference operator must introduce dissipation* (or amplification) into the modified PDE.

Note that the use of one-sided differencing schemes is not the only way to introduce dissipation. Any symmetric component in the spatial operator introduces dissipation (or amplification). Therefore, one could choose $\beta = 1/2$ in (11.2). The resulting spatial operator is not one-sided, but it is dissipative. Biased schemes use more information on one side of the node than the other. For example, a third-order backward-biased scheme is given by

$$(\delta_x u)_j = \frac{1}{6\Delta x}(u_{j-2} - 6u_{j-1} + 3u_j + 2u_{j+1})$$

$$= \frac{1}{12\Delta x}[(u_{j-2} - 8u_{j-1} + 8u_{j+1} - u_{j+2})$$

$$+(u_{j-2} - 4u_{j-1} + 6u_j - 4u_{j+1} + u_{j+2})] . \qquad (11.7)$$

The antisymmetric component of this operator is the fourth-order centered difference operator. The symmetric component approximates $\Delta x^3 u_{xxxx}/12$. Therefore, this operator produces fourth-order accuracy in phase with a third-order dissipative term.

11.3 The Lax–Wendroff Method

In order to introduce numerical dissipation using one-sided differencing, backward differencing must be used if the wave speed is positive, and forward differencing must be used if the wave speed is negative. Next we consider a method which introduces dissipation independent of the sign of the wave speed, known as the Lax–Wendroff method. This explicit method differs conceptually from the methods considered previously in which spatial differencing and time-marching are treated separately.

Consider the Taylor-series expansion in time

$$u(x, t+h) = u + h\frac{\partial u}{\partial t} + \frac{1}{2}h^2\frac{\partial^2 u}{\partial t^2} + O(h^3) . \qquad (11.8)$$

First replace the time derivatives with space derivatives according to the PDE (in this case, the linear convection equation $\frac{\partial u}{\partial t} + a\frac{\partial u}{\partial x} = 0$). Thus

$$\frac{\partial u}{\partial t} = -a\frac{\partial u}{\partial x}, \qquad \frac{\partial^2 u}{\partial t^2} = a^2\frac{\partial^2 u}{\partial x^2} . \qquad (11.9)$$

Now replace the space derivatives with three-point centered difference operators, giving

$$u_j^{(n+1)} = u_j^{(n)} - \frac{1}{2}\frac{ah}{\Delta x}(u_{j+1}^{(n)} - u_{j-1}^{(n)})$$

$$+ \frac{1}{2}\left(\frac{ah}{\Delta x}\right)^2 (u_{j+1}^{(n)} - 2u_j^{(n)} + u_{j-1}^{(n)}) . \tag{11.10}$$

This is the Lax–Wendroff method applied to the linear convection equation. It is a fully-discrete finite-difference scheme. There is no intermediate semi-discrete stage.

For periodic boundary conditions, the corresponding fully-discrete matrix operator is

$$\vec{u}_{n+1} = B_p \left(\frac{1}{2}\left[\frac{ah}{\Delta x} + \left(\frac{ah}{\Delta x}\right)^2\right], 1 - \left(\frac{ah}{\Delta x}\right)^2, \frac{1}{2}\left[-\frac{ah}{\Delta x} + \left(\frac{ah}{\Delta x}\right)^2\right]\right) \vec{u}_n .$$

The eigenvalues of this matrix are

$$\sigma_m = 1 - \left(\frac{ah}{\Delta x}\right)^2 \left[1 - \cos\left(\frac{2\pi m}{M}\right)\right] - \mathrm{i}\,\frac{ah}{\Delta x}\sin\left(\frac{2\pi m}{M}\right) ,$$

$$m = 0, 1, \ldots, M - 1 . \tag{11.11}$$

For $|\frac{ah}{\Delta x}| \leq 1$ all of the eigenvalues have modulus less than or equal to unity and hence the method is stable *independent of the sign of a*. The quantity $|\frac{ah}{\Delta x}|$ is known as the Courant (or CFL) number. It is equal to the ratio of the distance travelled by a wave in one time step to the mesh spacing.

The nature of the dissipative properties of the Lax–Wendroff scheme can be seen by examining the modified partial differential equation. This is derived by first substituting Taylor series expansions for all terms in (11.10) to obtain

$$\frac{\partial u}{\partial t} + a\frac{\partial u}{\partial x} + \frac{h}{2}\frac{\partial^2 u}{\partial t^2} + \frac{h^2}{6}\frac{\partial^3 u}{\partial t^3} + \frac{h^3}{24}\frac{\partial^4 u}{\partial t^4}$$

$$- \frac{ha^2}{2}\frac{\partial^2 u}{\partial x^2} + \frac{a\Delta x^2}{6}\frac{\partial^3 u}{\partial x^3} - \frac{ha^2\Delta x^2}{24}\frac{\partial^4 u}{\partial x^4} + \cdots = 0 . \tag{11.12}$$

The relations given in (11.9) cannot be used to convert the time derivatives in the error terms to space derivatives. Instead the modified PDE itself must be used. For example, applying the operator

$$-\frac{h}{2}\frac{\partial}{\partial t}$$

to (11.12) gives

$$-\frac{h}{2}\frac{\partial^2 u}{\partial t^2} - \frac{ha}{2}\frac{\partial^2 u}{\partial t\partial x} - \frac{h^2}{4}\frac{\partial^3 u}{\partial t^3} - \frac{h^3}{12}\frac{\partial^4 u}{\partial t^4}$$

$$+ \frac{h^2 a^2}{4}\frac{\partial^3 u}{\partial t\partial x^2} - \frac{ha\Delta x^2}{12}\frac{\partial^4 u}{\partial t\partial x^3} + \cdots = 0 . \tag{11.13}$$

Adding (11.13) to (11.12) eliminates the term

$$\frac{h}{2}\frac{\partial^2 u}{\partial t^2}$$

but introduces a new term

$$-\frac{ha}{2}\frac{\partial^2 u}{\partial t \partial x} .$$

This term can be eliminated by applying the operator

$$\frac{ha}{2}\frac{\partial}{\partial x}$$

to (11.12) and adding the result to the sum of (11.12) and (11.13). This process is repeated, eliminating one term at a time, until all of the time derivatives in the error terms have been replaced by space derivatives. The end result is the following modified partial differential equation for the Lax–Wendroff method:

$$\frac{\partial u}{\partial t} + a\frac{\partial u}{\partial x} = -\frac{a}{6}(\Delta x^2 - a^2 h^2)\frac{\partial^3 u}{\partial x^3} - \frac{a^2 h}{8}(\Delta x^2 - a^2 h^2)\frac{\partial^4 u}{\partial x^4} + \cdots .$$

The two leading error terms appear on the right side of the equation. Recall that the odd derivatives on the right side lead to unwanted dispersion and the even derivatives lead to dissipation (or amplification, depending on the sign). Therefore, the leading error term in the Lax–Wendroff method is dispersive and proportional to

$$-\frac{a}{6}(\Delta x^2 - a^2 h^2)\frac{\partial^3 u}{\partial x^3} = -\frac{a\Delta x^2}{6}(1 - C_n^2)\frac{\partial^3 u}{\partial x^3} ,$$

where $C_n = |\frac{ah}{\Delta x}|$ is the Courant number. The dissipative term is proportional to

$$-\frac{a^2 h}{8}(\Delta x^2 - a^2 h^2)\frac{\partial^4 u}{\partial x^4} = -\frac{a^2 h\Delta x^2}{8}(1 - C_n^2)\frac{\partial^4 u}{\partial x^4} .$$

This term has the appropriate sign and hence the scheme is truly dissipative as long as $C_n \leq 1$.

A closely related method is that of MacCormack. Recall MacCormack's time-marching method, presented in Chapter 6:

$$\tilde{u}_{n+1} = u_n + hu'_n$$

$$u_{n+1} = \frac{1}{2}[u_n + \tilde{u}_{n+1} + h\tilde{u}'_{n+1}] . \tag{11.14}$$

If we use first-order backward differencing in the first stage and first-order forward differencing in the second stage,[1] a dissipative second-order method is obtained. For the linear convection equation, this approach leads to

---

[1] Or vice versa; for nonlinear problems, these should be applied alternately.

$$\tilde{u}_j^{(n+1)} = u_j^{(n)} - \frac{ah}{\Delta x}(u_j^{(n)} - u_{j-1}^{(n)})$$

$$u_j^{(n+1)} = \frac{1}{2}[u_j^{(n)} + \tilde{u}_j^{(n+1)} - \frac{ah}{\Delta x}(\tilde{u}_{j+1}^{(n+1)} - \tilde{u}_j^{(n+1)})] , \qquad (11.15)$$

which can be shown to be identical to the Lax–Wendroff method. Hence MacCormack's method has the same dissipative and dispersive properties as the Lax–Wendroff method. The two methods differ when applied to nonlinear hyperbolic systems, however.

11.4 Upwind Schemes

In Section 11.1, we saw that numerical dissipation can be introduced in the spatial difference operator using one-sided difference schemes or, more generally, by adding a symmetric component to the spatial operator. With this approach, the direction of the one-sided operator (i.e. whether it is a forward or a backward difference) or the sign of the symmetric component depends on the sign of the wave speed. When a *hyperbolic system* of equations is being solved, the wave speeds can be both positive and negative. For example, the eigenvalues of the flux Jacobian for the one-dimensional Euler equations are $u, u + a, u - a$. When the flow is subsonic, these are of mixed sign. In order to apply one-sided differencing schemes to such systems, some form of splitting is required. This is avoided in the Lax–Wendroff scheme. However, as a result of their superior flexibility, schemes in which the numerical dissipation is introduced in the spatial operator are generally preferred over the Lax–Wendroff approach.

Consider again the linear convection equation:

$$\frac{\partial u}{\partial t} + a\frac{\partial u}{\partial x} = 0 , \qquad (11.16)$$

where we do not make any assumptions as to the sign of a. We can rewrite (11.16) as

$$\frac{\partial u}{\partial t} + (a^+ + a^-)\frac{\partial u}{\partial x} = 0 \quad , \quad a^\pm = \frac{a \pm |a|}{2} .$$

If $a \geq 0$, then $a^+ = a \geq 0$ and $a^- = 0$. Alternatively, if $a \leq 0$, then $a^+ = 0$ and $a^- = a \leq 0$. Now for the a^+ (≥ 0) term we can safely backward difference and for the a^- (≤ 0) term forward difference. This is the basic concept behind upwind methods, that is, some decomposition or splitting of the fluxes into terms which have positive and negative characteristic speeds so that appropriate differencing schemes can be chosen. In the next two sections, we present two splitting techniques commonly used with upwind methods. These are by no means unique.

The above approach to obtaining a stable discretization independent of the sign of a can be written in a different, but entirely equivalent, manner.

From (11.2), we see that a stable discretization is obtained with $\beta = 1$ if $a \geq 0$ and with $\beta = -1$ if $a \leq 0$. This is achieved by the following point operator:

$$-a(\delta_x u)_j = \frac{-1}{2\Delta x}[a(-u_{j-1} + u_{j+1}) + |a|(-u_{j-1} + 2u_j - u_{j+1})] \ .$$

$$(11.17)$$

This approach is extended to systems of equations in Section 11.5.

In this section, we present the basic ideas of flux-vector and flux-difference splitting. For more subtle aspects of implementation and application of such techniques to nonlinear hyperbolic systems such as the Euler equations, the reader is referred to the literature on this subject.

11.4.1 Flux-Vector Splitting

Recall from Section 2.5 that a linear, constant-coefficient, hyperbolic system of partial differential equations given by

$$\frac{\partial u}{\partial t} + \frac{\partial f}{\partial x} = \frac{\partial u}{\partial t} + A\frac{\partial u}{\partial x} = 0, \qquad (11.18)$$

where $f = Au$, can be decoupled into characteristic equations of the form

$$\frac{\partial w_i}{\partial t} + \lambda_i \frac{\partial w_i}{\partial x} = 0 \ . \qquad (11.19)$$

where the wave speeds, λ_i, are the eigenvalues of the Jacobian matrix, A, and the w_i's are the characteristic variables. In order to apply a one-sided (or biased) spatial differencing scheme, we need to apply a backward difference if the wave speed, λ_i, is positive, and a forward difference if the wave speed is negative. To accomplish this, let us split the matrix of eigenvalues, Λ, into two components such that

$$\Lambda = \Lambda^+ + \Lambda^- \ , \qquad (11.20)$$

where

$$\Lambda^+ = \frac{\Lambda + |\Lambda|}{2}, \qquad \Lambda^- = \frac{\Lambda - |\Lambda|}{2} \ . \qquad (11.21)$$

With these definitions, Λ^+ contains the positive eigenvalues and Λ^- contains the negative eigenvalues. We can now rewrite the system in terms of characteristic variables as

$$\frac{\partial w}{\partial t} + \Lambda\frac{\partial w}{\partial x} = \frac{\partial w}{\partial t} + \Lambda^+\frac{\partial w}{\partial x} + \Lambda^-\frac{\partial w}{\partial x} = 0 \ . \qquad (11.22)$$

The spatial terms have been split into two components according to the sign of the wave speeds. We can use backward differencing for the $\Lambda^+ \frac{\partial w}{\partial x}$ term and forward differencing for the $\Lambda^- \frac{\partial w}{\partial x}$ term. Premultiplying by X, the matrix of right eigenvectors of A, and inserting the product $X^{-1}X$ in the spatial terms gives

$$\frac{\partial Xw}{\partial t} + \frac{\partial X\Lambda^+ X^{-1}Xw}{\partial x} + \frac{\partial X\Lambda^- X^{-1}Xw}{\partial x} = 0 \ . \tag{11.23}$$

With the definitions[2]

$$A^+ = X\Lambda^+ X^{-1}, \qquad A^- = X\Lambda^- X^{-1} \tag{11.24}$$

and recalling that $u = Xw$, we obtain

$$\frac{\partial u}{\partial t} + \frac{\partial A^+ u}{\partial x} + \frac{\partial A^- u}{\partial x} = 0 \ . \tag{11.25}$$

Finally the split flux vectors are defined as

$$f^+ = A^+ u, \qquad f^- = A^- u \tag{11.26}$$

and we can write

$$\frac{\partial u}{\partial t} + \frac{\partial f^+}{\partial x} + \frac{\partial f^-}{\partial x} = 0 \ . \tag{11.27}$$

In the linear case, the definition of the split fluxes follows directly from the definition of the flux, $f = Au$. For the Euler equations, f is also equal to Au as a result of their homogeneous property, as discussed in Appendix C. Note that

$$f = f^+ + f^- \ . \tag{11.28}$$

Thus by applying backward differences to the f^+ term and forward differences to the f^- term, we are in effect solving the characteristic equations in the desired manner. This approach is known as flux-vector splitting.

When an implicit time-marching method is used, the Jacobians of the split flux vectors are required. In the nonlinear case,

$$\frac{\partial f^+}{\partial u} \neq A^+, \qquad \frac{\partial f^-}{\partial u} \neq A^- \ . \tag{11.29}$$

Therefore, one must find and use the new Jacobians given by

$$A^{++} = \frac{\partial f^+}{\partial u}, \qquad A^{--} = \frac{\partial f^-}{\partial u} \ . \tag{11.30}$$

For the Euler equations, A^{++} has eigenvalues which are all positive, and A^{--} has all negative eigenvalues.

---

[2] With these definitions, A^+ has all positive eigenvalues, and A^- has all negative eigenvalues.

11.4.2 Flux-Difference Splitting

Another approach, more suited to finite-volume methods, is known as flux-difference splitting.[3] In a finite-volume method, the fluxes must be evaluated at cell boundaries. We again begin with the diagonalized form of the linear, constant-coefficient, hyperbolic system of equations

$$\frac{\partial w}{\partial t} + \Lambda \frac{\partial w}{\partial x} = 0 . \tag{11.31}$$

The flux vector associated with this form is $g = \Lambda w$. Now, as in Chapter 5, we consider the numerical flux at the interface between nodes j and $j+1$, $\hat{g}_{j+1/2}$, as a function of the states to the left and right of the interface, w^L and w^R, respectively. The centered approximation to $g_{j+1/2}$, which is nondissipative, is given by

$$\hat{g}_{j+1/2} = \frac{1}{2}[g(w^L) + g(w^R)] . \tag{11.32}$$

In order to obtain a one-sided *upwind* approximation, we require

$$(\hat{g}_i)_{j+1/2} = \begin{cases} \lambda_i(w_i)^L \text{ if } \lambda_i > 0 \\ \lambda_i(w_i)^R \text{ if } \lambda_i < 0 \end{cases} . \tag{11.33}$$

where the subscript i indicates individual components of w and g. This is achieved with

$$(\hat{g}_i)_{j+1/2} = \frac{1}{2}\lambda_i \left[(w_i)^L + (w_i)^R \right] + \frac{1}{2}|\lambda_i| \left[(w_i)^L - (w_i)^R \right] \tag{11.34}$$

or

$$\hat{g}_{j+1/2} = \frac{1}{2}\Lambda \left(w^L + w^R \right) + \frac{1}{2}|\Lambda| \left(w^L - w^R \right) . \tag{11.35}$$

Now, as in (11.23), we premultiply by X to return to the original variables and insert the product $X^{-1}X$ after Λ and $|\Lambda|$ to obtain

$$X\hat{g}_{j+1/2} = \frac{1}{2}X\Lambda X^{-1}X \left(w^L + w^R \right) + \frac{1}{2}X|\Lambda|X^{-1}X \left(w^L - w^R \right) \tag{11.36}$$

and thus

$$\hat{f}_{j+1/2} = \frac{1}{2} \left(f^L + f^R \right) + \frac{1}{2}|A| \left(u^L - u^R \right) , \tag{11.37}$$

where

---

[3] For nonlinear systems, there are subtle but important differences between flux-vector and flux-difference splitting which we do not discuss here; the reader is advised to consult the literature.

$$|A| = X|\Lambda|X^{-1} \tag{11.38}$$

and we have also used the relations $f = Xg = Au$, $u = Xw$, and $A = X\Lambda X^{-1}$.

In the linear, constant-coefficient case, this leads to an upwind operator which is identical to that obtained using flux-vector splitting. However, in the nonlinear case, there is some ambiguity regarding the definition of $|A|$ at the cell interface $j + 1/2$. In order to resolve this, consider a situation in which the eigenvalues of A are all of the same sign. In this case, we would like our definition of $\hat{f}_{j+1/2}$ to satisfy

$$\hat{f}_{j+1/2} = \begin{cases} f^L & \text{if all } \lambda_i\text{'s} > 0 \\ f^R & \text{if all } \lambda_i\text{'s} < 0 \end{cases} \tag{11.39}$$

giving pure upwinding. If the eigenvalues of A are all positive, $|A| = A$; if they are all negative, $|A| = -A$. Hence satisfaction of (11.39) is obtained by the definition

$$\hat{f}_{j+1/2} = \frac{1}{2}\left(f^L + f^R\right) + \frac{1}{2}|A_{j+1/2}|\left(u^L - u^R\right) \tag{11.40}$$

if $A_{j+1/2}$ satisfies

$$f^L - f^R = A_{j+1/2}\left(u^L - u^R\right) . \tag{11.41}$$

For the Euler equations for a perfect gas, (11.41) is satisfied by the flux Jacobian evaluated at the Roe-average state given by

$$u_{j+1/2} = \frac{\sqrt{\rho^L}u^L + \sqrt{\rho^R}u^R}{\sqrt{\rho^L} + \sqrt{\rho^R}} \tag{11.42}$$

$$H_{j+1/2} = \frac{\sqrt{\rho^L}H^L + \sqrt{\rho^R}H^R}{\sqrt{\rho^L} + \sqrt{\rho^R}} \tag{11.43}$$

where u and $H = (e + p)/\rho$ are the velocity and the total enthalpy per unit mass, respectively.[4]

11.5 Artificial Dissipation

We have seen that numerical dissipation can be introduced by using one-sided differencing schemes together with some form of flux splitting. We have also seen that such dissipation can be introduced by adding a symmetric component to an antisymmetric (dissipation-free) operator. Thus we can generalize

---

[4] Note that the flux Jacobian can be written in terms of u and H only; see Exercise 11.6.

the concept of upwinding to include any scheme in which the symmetric portion of the operator is treated in such a manner as to be truly dissipative.

For example, let

$$(\delta_x^a u)_j = \frac{u_{j+1} - u_{j-1}}{2\Delta x}, \qquad (\delta_x^s u)_j = \frac{-u_{j+1} + 2u_j - u_{j-1}}{2\Delta x} . \qquad (11.44)$$

Applying $\delta_x = \delta_x^a + \delta_x^s$ to the spatial derivative in (11.19) is stable if $\lambda_i \geq 0$ and unstable if $\lambda_i < 0$. Similarly, applying $\delta_x = \delta_x^a - \delta_x^s$ is stable if $\lambda_i \leq 0$ and unstable if $\lambda_i > 0$. The appropriate implementation is thus

$$\lambda_i \delta_x = \lambda_i \delta_x^a + |\lambda_i| \delta_x^s . \qquad (11.45)$$

Extension to a hyperbolic system by applying the above approach to the characteristic variables, as in the previous two sections, gives

$$\delta_x(Au) = \delta_x^a(Au) + \delta_x^s(|A|u) \qquad (11.46)$$

or

$$\delta_x f = \delta_x^a f + \delta_x^s(|A|u) , \qquad (11.47)$$

where $|A|$ is defined in (11.38). The second spatial term is known as *artificial dissipation*. It is also sometimes referred to as artificial diffusion or artificial viscosity. With appropriate choices of δ_x^a and δ_x^s, this approach can be related to the upwind approach. This is particularly evident from a comparison of (11.40) and (11.47).

It is common to use the following operator for δ_x^s

$$(\delta_x^s u)_j = \frac{\epsilon}{\Delta x}(u_{j-2} - 4u_{j-1} + 6u_j - 4u_{j+1} + u_{j+2}) , \qquad (11.48)$$

where ϵ is a problem-dependent coefficient. This symmetric operator approximates $\epsilon \Delta x^3 u_{xxxx}$ and thus introduces a third-order dissipative term. With an appropriate value of ϵ, this often provides sufficient damping of high frequency modes without greatly affecting the low frequency modes. For details of how this can be implemented for nonlinear hyperbolic systems, the reader should consult the literature. A more complicated treatment of the numerical dissipation is also required near shock waves and other discontinuities, but is beyond the scope of this book.

Exercises

11.1 A second-order backward difference approximation to a first derivative is given as a point operator by

$$(\delta_x u)_j = \frac{1}{2\Delta x}(u_{j-2} - 4u_{j-1} + 3u_j) .$$

(a) Express this operator in banded matrix form (for periodic boundary conditions), then derive the symmetric and skew-symmetric matrices that have the matrix operator as their sum. (See Appendix A.3 to see how to construct the symmetric and skew-symmetric components of a matrix.)

(b) Using a Taylor table, find the derivative which is approximated by the corresponding symmetric and skew-symmetric operators and the leading error term for each.

11.2 Find the modified wavenumber for the first-order backward difference operator. Plot the real and imaginary parts of $\kappa^* \Delta x$ vs. $\kappa \Delta x$ for $0 \leq \kappa \Delta x \leq \pi$. Using Fourier analysis as in Section 6.6.2, find $|\sigma|$ for the combination of this spatial operator with 4th-order Runge–Kutta time marching at a Courant number of unity and plot vs. $\kappa \Delta x$ for $0 \leq \kappa \Delta x \leq \pi$.

11.3 Find the modified wavenumber for the operator given in (11.7). Plot the real and imaginary parts of $\kappa^* \Delta x$ vs. $\kappa \Delta x$ for $0 \leq \kappa \Delta x \leq \pi$. Using Fourier analysis as in Section 6.6.2, find $|\sigma|$ for the combination of this spatial operator with 4th-order Runge–Kutta time marching at a Courant number of unity and plot vs. $\kappa \Delta x$ for $0 \leq \kappa \Delta x \leq \pi$.

11.4 Consider the spatial operator obtained by combining second-order centered differences with the symmetric operator given in (11.48). Find the modified wavenumber for this operator with $\epsilon = 0, 1/12, 1/24$, and $1/48$. Plot the real and imaginary parts of $\kappa^* \Delta x$ vs. $\kappa \Delta x$ for $0 \leq \kappa \Delta x \leq \pi$. Using Fourier analysis as in Section 6.6.2, find $|\sigma|$ for the combination of this spatial operator with 4th-order Runge–Kutta time marching at a Courant number of unity and plot vs. $\kappa \Delta x$ for $0 \leq \kappa \Delta x \leq \pi$.

11.5 Consider the hyperbolic system derived in Exercise 2.8. Find the matrix $|A|$. Form the plus-minus split flux vectors as in Section 11.4.1.

11.6 Show that the flux Jacobian for the 1D Euler equations can be written in terms of u and H. Show that the use of the Roe average state given in (11.42) and (11.43) leads to satisfaction of (11.41).

12. Split and Factored Forms

In the next two chapters, we present and analyze split and factored algorithms. This gives the reader a feel for some of the modifications which can be made to the basic algorithms in order to obtain efficient solvers for practical multidimensional applications, and a means for analyzing such modified forms.

12.1 The Concept

Factored forms of numerical operators are used extensively in constructing and applying numerical methods to problems in fluid mechanics. They are the basis for a wide variety of methods variously known by the labels "hybrid", "time split", and "fractional step". Factored forms are especially useful for the derivation of practical algorithms that use implicit methods. When we approach numerical analysis in the light of matrix derivative operators, the concept of factoring is quite simple to present and grasp. Let us start with the following observations:

(1) Matrices can be split in quite arbitrary ways.
(2) Advancing to the next time level always requires some reference to a previous one.
(3) Time-marching methods are valid only to some order of accuracy in the step size, h.

Now recall the generic ODE's produced by the semi-discrete approach

$$\frac{d\vec{u}}{dt} = A\vec{u} - \vec{f} \tag{12.1}$$

and consider the above observations. From observation (1) (arbitrary splitting of A) :

$$\frac{d\vec{u}}{dt} = [A_1 + A_2]\vec{u} - \vec{f}, \tag{12.2}$$

where $A = [A_1 + A_2]$, but A_1 and A_2 are not unique. For the time march let us choose the simple, first-order[1], explicit Euler method. Then, from observation

---
[1] Second-order time-marching methods are considered later.

(2) (new data $\vec{u}_{n+1}$ in terms of old $\vec{u}_n$):

$$\vec{u}_{n+1} = [\, I \, + hA_1 + hA_2]\vec{u}_n - h\vec{f} + O(h^2) \tag{12.3}$$

or its equivalent

$$\vec{u}_{n+1} = \big[[\, I \, + hA_1][\, I \, + hA_2] - h^2 A_1 A_2\big]\vec{u}_n - h\vec{f} + O(h^2) \ .$$

Finally, from observation (3) (allowing us to drop higher order terms):

$$\vec{u}_{n+1} = [\, I \, + hA_1][\, I \, + hA_2]\vec{u}_n - h\vec{f} + O(h^2) \ . \tag{12.4}$$

Notice that (12.3) and (12.4) have the same formal order of accuracy and, in this sense, neither one is to be preferred over the other. However, their numerical stability can be quite different, and techniques to carry out their numerical evaluation can have arithmetic operation counts that vary by orders of magnitude. Both of these considerations are investigated later. Here we seek only to apply to some simple cases the concept of factoring.

12.2 Factoring Physical Representations – Time Splitting

Suppose we have a PDE that represents both the processes of convection and dissipation. The semi-discrete approach to its solution might be put in the form

$$\frac{d\vec{u}}{dt} = A_c \vec{u} + A_d \vec{u} + (\vec{bc}) \ , \tag{12.5}$$

where A_c and A_d are matrices representing the convection and dissipation terms, respectively, and their sum forms the A matrix we have considered in the previous sections. Choose again the explicit Euler time march so that

$$\vec{u}_{n+1} = [\, I \, + hA_d + hA_c]\vec{u}_n + h(\vec{bc}) + O(h^2) \ . \tag{12.6}$$

Now consider the factored form

$$\vec{u}_{n+1} = [\, I \, + hA_d]\Big([\, I \, + hA_c]\vec{u}_n + h(\vec{bc})\Big)$$

$$= \underbrace{[\, I \, + hA_d + hA_c]\vec{u}_n + h(\vec{bc})}_{\text{original unfactored terms}}$$

$$+ \underbrace{h^2 A_d \Big(A_c \vec{u}_n + (\vec{bc})\Big)}_{\text{higher-order terms}} + O(h^2) \tag{12.7}$$

and we see that (12.7) and the original unfactored form (12.6) have identical orders of accuracy in the time approximation. Therefore, on this basis, their selection is arbitrary. In practical applications[2] equations such as (12.7) are often applied in a predictor–corrector sequence. In this case one could write

$$\tilde{u}_{n+1} = [\,I\,+ hA_{\mathrm{c}}]\vec{u}_n + h(\vec{bc})$$
$$\vec{u}_{n+1} = [\,I\,+ hA_{\mathrm{d}}]\tilde{u}_{n+1}\,. \tag{12.8}$$

Factoring can also be useful to form split combinations of implicit and explicit techniques. For example, another way to approximate (12.5) with the same order of accuracy is given by the expression

$$\vec{u}_{n+1} = [\,I\,- hA_{\mathrm{d}}]^{-1}\Big([\,I\,+ hA_{\mathrm{c}}]\vec{u}_n + h(\vec{bc})\Big)$$
$$= \underbrace{[\,I\,+ hA_{\mathrm{d}} + hA_{\mathrm{c}}]\vec{u}_n + h(\vec{bc})}_{\text{original unfactored terms}} + O(h^2)\,, \tag{12.9}$$

where in this approximation we have used the fact that

$$[\,I\,- hA_{\mathrm{d}}]^{-1} = I + hA_{\mathrm{d}} + h^2 A_{\mathrm{d}}^2 + \cdots$$

if $h \cdot \|A_{\mathrm{d}}\| < 1$, where $\|A_{\mathrm{d}}\|$ is some norm of $[A_{\mathrm{d}}]$. This time a predictor–corrector interpretation leads to the sequence

$$\tilde{u}_{n+1} = [\,I\,+ hA_{\mathrm{c}}]\vec{u}_n + h(\vec{bc})$$
$$[\,I\,- hA_{\mathrm{d}}]\vec{u}_{n+1} = \tilde{u}_{n+1}\,. \tag{12.10}$$

The convection operator is applied explicitly, as before, but the diffusion operator is now implicit, requiring a tridiagonal solver if the diffusion term is central differenced. Since numerical stiffness is generally much more severe for the diffusion process, this factored form would appear to be superior to that provided by (12.8). However, the important aspect of stability has yet to be discussed.

We should mention here that (12.9) can be derived for a different point of view by writing (12.6) in the form

$$\frac{u_{n+1} - u_n}{h} = A_{\mathrm{c}}u_n + A_{\mathrm{d}}u_{n+1} + (\vec{bc}) + O(h^2)\,.$$

Then

$$[\,I\,- hA_{\mathrm{d}}]u_{n+1} = [\,I\,+ hA_{\mathrm{c}}]u_n + h(\vec{bc})\,,$$

which is identical to (12.10).

---

[2] We do not suggest that this particular method is suitable for use. We have yet to determine its stability, and a first-order time-march method is usually unsatisfactory.

12.3 Factoring Space Matrix Operators in 2D

12.3.1 Mesh Indexing Convention

Factoring is widely used in codes designed for the numerical solution of equations governing unsteady two- and three-dimensional flows. Let us study the basic concept of factoring by inspecting its use on the linear 2D scalar PDE that models diffusion

$$\frac{\partial u}{\partial t} = \frac{\partial^2 u}{\partial x^2} + \frac{\partial^2 u}{\partial y^2} . \tag{12.11}$$

We begin by reducing this PDE to a coupled set of ODE's by differencing the space derivatives and inspecting the resulting matrix operator.

A clear description of a matrix finite-difference operator in two and three dimensions requires some reference to a mesh. We choose the 3×4 point mesh[3] shown below. In this example M_x, the number of (interior) x points, is 4 and M_y, the number of (interior) y points is 3. The numbers 11, 12, $\cdots$, 43 represent the location in the mesh of the dependent variable bearing that index. Thus u_{32} represents the value of u at $j = 3$ and $k = 2$.

$$
\begin{array}{ccccccc}
 & & \odot & \odot & \odot & \odot & \\
M_y & \odot & 13 & 23 & 33 & 43 & \odot \\
k & \odot & 12 & 22 & 32 & 42 & \odot \\
1 & \odot & 11 & 21 & 31 & 41 & \odot \\
 & & \odot & \odot & \odot & \odot & \\
 & & 1 & j & \cdots & M_x &
\end{array}
$$

Mesh indexing in 2D

12.3.2 Data-Bases and Space Vectors

The dimensioned array in a computer code that allots the storage locations of the dependent variable(s) is referred to as a *data-base*. There are many ways to lay out a data-base. Of these, we consider only two: (1) consecutively along rows that are themselves consecutive from $k = 1$ to M_y, and (2) consecutively along columns that are consecutive from $j = 1$ to M_x. We refer to each row or column group as a *space vector* (they represent data along lines that are continuous in space) and label their sum with the symbol U. In particular, (1) and (2) above are referred to as x-vectors and y-vectors, respectively. The

---
[3] This could also be called a 5×6 point mesh if the boundary points (labeled $\odot$ in the sketch) were included, but in these notes we describe the size of a mesh by the number of *interior* points.

symbol U by itself is not enough to identify the structure of the data-base and is used only when the structure is immaterial or understood.

To be specific about the structure, we label a data-base composed of x-vectors with $U^{(x)}$, and one composed of y-vectors with $U^{(y)}$. Examples of the order of indexing for these space vectors are given in (12.17a) and (12.17b).

12.3.3 Data-Base Permutations

The two vectors (arrays) are related by a permutation matrix P such that

$$U^{(x)} = P_{xy} U^{(y)} \quad \text{and} \quad U^{(y)} = P_{yx} U^{(x)} , \tag{12.12}$$

where

$$P_{yx} = P_{xy}^{\mathrm{T}} = P_{xy}^{-1} .$$

Now consider the structure of a matrix finite-difference operator representing 3-point central-differencing schemes for both space derivatives in two dimensions. When the matrix is multiplying a space vector U, the usual (but ambiguous) representation is given by A_{x+y}. In this notation the ODE form of (12.11) can be written[4]

$$\frac{\mathrm{d}U}{\mathrm{d}t} = A_{x+y}U + (\vec{bc}) . \tag{12.13}$$

If it is important to be specific about the data-base structure, we use the notation $A_{x+y}^{(x)}$ or $A_{x+y}^{(y)}$, depending on the data-base chosen for the U it multiplies. Examples are in (12.17). Notice that the matrices are not the same although they represent the same derivative operation. Their structures are similar, however, and they are related by the same permutation matrix that relates $U^{(x)}$ to $U^{(y)}$. Thus

$$A_{x+y}^{(x)} = P_{xy} \cdot A_{x+y}^{(y)} \cdot P_{yx} . \tag{12.14}$$

12.3.4 Space Splitting and Factoring

We are now prepared to discuss splitting in two dimensions. It should be clear that the matrix $A_{x+y}^{(x)}$ can be split into two matrices such that

$$A_{x+y}^{(x)} = A_x^{(x)} + A_y^{(x)} , \tag{12.15}$$

where $A_x^{(x)}$ and $A_y^{(x)}$ are shown in (12.18). Similarily

$$A_{x+y}^{(y)} = A_x^{(y)} + A_y^{(y)} , \tag{12.16}$$

---

[4] Notice that A_{x+y} and U, which are notations used in the special case of space vectors, are subsets of A and $\vec{u}$, used in the previous sections.

$$A^{(x)}_{x+y} \cdot U^{(x)} = \left[\begin{array}{cccc|cccc|cccc}
\bullet & x & & & o & & & & & & & \\
x & \bullet & x & & & o & & & & & & \\
 & x & \bullet & x & & & o & & & & & \\
 & & x & \bullet & & & & o & & & & \\
\hline
o & & & & \bullet & x & & & o & & & \\
 & o & & & x & \bullet & x & & & o & & \\
 & & o & & & x & \bullet & x & & & o & \\
 & & & o & & & x & \bullet & & & & o \\
\hline
 & & & & o & & & & \bullet & x & & \\
 & & & & & o & & & x & \bullet & x & \\
 & & & & & & o & & & x & \bullet & x \\
 & & & & & & & o & & & x & \bullet
\end{array}\right]
\begin{array}{l} 11\\21\\31\\41\\12\\22\\32\\42\\13\\23\\33\\43 \end{array}$$

(a) Elements in two-dimensional, central-difference, matrix operator, A_{x+y}, for 3×4 mesh shown in Section 12.3.1. Data-base composed of M_y x-vectors stored in $U^{(x)}$. Entries for $x \to x$, for $y \to o$, for both $\to \bullet$

(12.17)

$$A^{(y)}_{x+y} \cdot U^{(y)} = \left[\begin{array}{ccc|ccc|ccc|ccc}
\bullet & o & & x & & & & & & & & \\
o & \bullet & o & & x & & & & & & & \\
 & o & \bullet & & & x & & & & & & \\
\hline
x & & & \bullet & o & & x & & & & & \\
 & x & & o & \bullet & o & & x & & & & \\
 & & x & & o & \bullet & & & x & & & \\
\hline
 & & & x & & & \bullet & o & & x & & \\
 & & & & x & & o & \bullet & o & & x & \\
 & & & & & x & & o & \bullet & & & x \\
\hline
 & & & & & & x & & & \bullet & o & \\
 & & & & & & & x & & o & \bullet & o \\
 & & & & & & & & x & & o & \bullet
\end{array}\right]
\begin{array}{l} 11\\12\\13\\21\\22\\23\\31\\32\\33\\41\\42\\43 \end{array}$$

(b) Elements in two-dimensional, central-difference, matrix operator, A_{x+y}, for 3×4 mesh shown in Section 12.3.1. Data-base composed of M_x y-vectors stored in $U^{(y)}$. Entries for $x \to x$, for $y \to o$, for both $\to \bullet$

$$
A_x^{(x)} \cdot U^{(x)} =
\begin{bmatrix}
x & x & & & & & & & & \\
x & x & x & & & & & & & \\
 & x & x & x & & & & & & \\
 & & x & x & & & & & & \\
 & & & & x & x & & & & \\
 & & & & x & x & x & & & \\
 & & & & & x & x & x & & \\
 & & & & & & x & x & & \\
 & & & & & & & & x & x \\
 & & & & & & & & x & x & x \\
 & & & & & & & & & x & x & x \\
 & & & & & & & & & & x & x
\end{bmatrix}
\cdot U^{(x)}
$$

$$
A_y^{(x)} \cdot U^{(x)} =
\begin{bmatrix}
o & & & & o & & & & & \\
 & o & & & & o & & & & \\
 & & o & & & & o & & & \\
 & & & o & & & & o & & \\
o & & & & o & & & & o & \\
 & o & & & & o & & & & o \\
 & & o & & & & o & & & & o \\
 & & & o & & & & o & & & & o \\
 & & & & o & & & & o & \\
 & & & & & o & & & & o \\
 & & & & & & o & & & & o \\
 & & & & & & & o & & & & o
\end{bmatrix}
\cdot U^{(x)}
$$

(12.18)

The splitting of $A_{x+y}^{(x)}$

where the split matrices are shown in (12.19).

The permutation relation also holds for the split matrices, so

$$
A_y^{(x)} = P_{xy} A_y^{(y)} P_{yx}
$$

and

$$
A_x^{(x)} = P_{xy} A_x^{(y)} P_{yx} .
$$

$$
A_x^{(y)} \cdot U^{(y)} =
\begin{bmatrix}
x & & & \vline & x & & & \vline & & & & \vline & & & \\
& x & & \vline & & x & & \vline & & & & \vline & & & \\
& & x & \vline & & & x & \vline & & & & \vline & & & \\
\hline
x & & & \vline & x & & & \vline & x & & & \vline & & & \\
& x & & \vline & & x & & \vline & & x & & \vline & & & \\
& & x & \vline & & & x & \vline & & & x & \vline & & & \\
\hline
& & & \vline & x & & & \vline & x & & & \vline & x & & \\
& & & \vline & & x & & \vline & & x & & \vline & & x & \\
& & & \vline & & & x & \vline & & & x & \vline & & & x \\
\hline
& & & \vline & & & & \vline & x & & & \vline & x & & \\
& & & \vline & & & & \vline & & x & & \vline & & x & \\
& & & \vline & & & & \vline & & & x & \vline & & & x
\end{bmatrix}
\cdot U^{(y)}
$$

(12.19)

$$
A_y^{(y)} \cdot U^{(y)} =
\begin{bmatrix}
o & o & & \vline & & & \vline & & \\
o & o & o & \vline & & & \vline & & \\
& o & o & \vline & & & \vline & & \\
\hline
& & & \vline & o & o & & \vline & \\
& & & \vline & o & o & o & \vline & \\
& & & \vline & & o & o & \vline & \\
\hline
& & & \vline & & & \vline & o & o \\
& & & \vline & & & \vline & o & o & o \\
& & & \vline & & & \vline & & o & o
\end{bmatrix}
\cdot U^{(y)}
$$

The splitting of $A_{x+y}^{(y)}$

The splittings in (12.15) and (12.16) can be combined with factoring in the manner described in Section 12.2. As an example (first-order in time), applying the implicit Euler method to (12.13) gives

$$
U_{n+1}^{(x)} = U_n^{(x)} + h\left[A_x^{(x)} + A_y^{(x)}\right]U_{n+1}^{(x)} + h(\vec{bc})
$$

or

$$
\left[I - hA_x^{(x)} - hA_y^{(x)}\right]U_{n+1}^{(x)} = U_n^{(x)} + h(\vec{bc}) + O(h^2) .
$$

(12.20)

As in Section 12.2, we retain the same first-order accuracy with the alternative

$$\left[I - hA_x^{(x)}\right]\left[I - hA_y^{(x)}\right]U_{n+1}^{(x)} = U_n^{(x)} + h(\vec{bc}) + O(h^2) . \qquad (12.21)$$

Write this in predictor–corrector form and permute the data-base of the second row. The result is

$$\left[I - hA_x^{(x)}\right]\tilde{U}^{(x)} = U_n^{(x)} + h(\vec{bc})$$

$$\left[I - hA_y^{(y)}\right]U_{n+1}^{(y)} = \tilde{U}^{(y)} . \qquad (12.22)$$

12.4 Second-Order Factored Implicit Methods

Second-order accuracy in time can be maintained in certain factored implicit methods. For example, apply the trapezoidal method to (12.13) where the derivative operators have been split as in (12.15) or (12.16). Let the data-base be immaterial and the $(\vec{bc})$ be time invariant. There results

$$\left[I - \frac{1}{2}hA_x - \frac{1}{2}hA_y\right]U_{n+1} = \left[I + \frac{1}{2}hA_x + \frac{1}{2}hA_y\right]U_n + h(\vec{bc}) + O(h^3) .$$

$$(12.23)$$

Factor both sides giving

$$\left[\left[I - \frac{1}{2}hA_x\right]\left[I - \frac{1}{2}hA_y\right] - \frac{1}{4}h^2 A_x A_y\right]U_{n+1}$$

$$= \left[\left[I + \frac{1}{2}hA_x\right]\left[I + \frac{1}{2}hA_y\right] - \frac{1}{4}h^2 A_x A_y\right]U_n + h(\vec{bc}) + O(h^3) .$$

$$(12.24)$$

Then notice that the combination $\frac{1}{4}h^2[A_x A_y](U_{n+1} - U_n)$ is proportional to h^3 since the leading term in the expansion of $(U_{n+1} - U_n)$ is proportional to h. Therefore, we can write

$$\left[I - \frac{1}{2}hA_x\right]\left[I - \frac{1}{2}hA_y\right]U_{n+1} = \left[I + \frac{1}{2}hA_x\right]\left[I + \frac{1}{2}hA_y\right]U_n$$

$$+h(\vec{bc}) + O(h^3) \qquad (12.25)$$

and both the factored and unfactored form of the trapezoidal method are second-order accurate in the time march.

An alternative form of this kind of factorization is the classical ADI (alternating direction implicit) method[5] usually written

---

[5] A form of the Douglas or Peaceman–Rachford methods.

$$\left[I - \frac{1}{2}hA_x\right]\tilde{U} = \left[I + \frac{1}{2}hA_y\right]U_n + \frac{1}{2}hF_n$$

$$\left[I - \frac{1}{2}hA_y\right]U_{n+1} = \left[I + \frac{1}{2}hA_x\right]\tilde{U} + \frac{1}{2}hF_{n+1} + O(h^3) . \quad (12.26)$$

For idealized commuting systems the methods given by (12.25) and (12.26) differ only in their evaluation of a time-dependent forcing term.

12.5 Importance of Factored Forms in Two and Three Dimensions

When the time-march equations are stiff, and implicit methods are required to permit reasonably large time steps, the use of factored forms becomes a very valuable tool for realistic problems. Consider, for example, the problem of computing the time advance in the unfactored form of the trapezoidal method given by (12.23)

$$\left[I - \frac{1}{2}hA_{x+y}\right]U_{n+1} = \left[I + \frac{1}{2}hA_{x+y}\right]U_n + h(\vec{bc}) .$$

Forming the right-hand side poses no problem, but finding U_{n+1} requires the solution of a sparse, but very large, set of coupled simultaneous equations having the matrix form shown in (12.17a) or (12.17b). Furthermore, in real cases involving the Euler or Navier–Stokes equations, each symbol $(o, x, \bullet)$ represents a 4×4 block matrix with entries that depend on the pressure, density and velocity field. Suppose we were to solve the equations directly. The forward sweep of a simple Gaussian elimination fills[6] all of the 4×4 blocks between the main and outermost diagonal[7] (e.g. between $\bullet$ and o in (12.17b)). This must be stored in computer memory to be used to find the final solution in the backward sweep. If N_e represents the order of the small block matrix (4 in the 2D Euler case), the approximate memory requirement is

$$(N_e \times M_y) \cdot (N_e \times M_y) \cdot M_x$$

floating point words. Here it is assumed that $M_y < M_x$. If $M_y > M_x$, M_y and M_x would be interchanged. A moderate mesh of 60×200 points would require over 11 million words to find the solution. Actually current computer power is able to cope rather easily with storage requirements of this order of magnitude. With computing speeds of over one gigaflop[8], direct solvers may

---

[6] For matrices as small as those shown there are many gaps in this "fill", but for meshes of practical size the fill is mostly dense.

[7] The lower band is also computed but does not have to be saved unless the solution is to be repeated for another vector.

[8] One billion floating-point operations per second.

become useful for finding steady-state solutions of practical problems in two dimensions. However, a three-dimensional solver would require a memory of approximately

$$N_e^2 \cdot M_y^2 \cdot M_z^2 \cdot M_x$$

words, and, for well resolved flow fields, this probably exceeds memory availability for some time to come.

On the other hand, consider computing a solution using the *factored* implicit equation (12.24). Again computing the right-hand side poses no problem. Accumulate the result of such a computation in the array (RHS). One can then write the remaining terms in the two-step predictor–corrector form

$$\left[I - \frac{1}{2}hA_x^{(x)}\right]\tilde{U}^{(x)} = (RHS)^{(x)}$$

$$\left[I - \frac{1}{2}hA_y^{(y)}\right]U_{n+1}^{(y)} = \tilde{U}^{(y)} , \qquad (12.27)$$

which has the same appearance as (12.22) but is second-order time accurate. The first step would be solved using M_y uncoupled block tridiagonal solvers[9]. Inspecting the top of (12.18), we see that this is equivalent to solving M_y *one-dimensional* problems, each with M_x blocks of order N_e. The temporary solution $\tilde{U}^{(x)}$ would then be permuted to $\tilde{U}^{(y)}$, and an inspection of the bottom of (12.19) shows that the final step consists of solving M_x *one-dimensional* implicit problems each with dimension M_y.

12.6 The Delta Form

Clearly many ways can be devised to split the matrices and generate factored forms. One way that is especially useful for ensuring a correct steady-state solution in a converged time-march is referred to as the "delta form", and we develop it next.

Consider the unfactored form of the trapezoidal method given by (12.23) and let the $(\vec{bc})$ be time invariant:

$$\left[I - \frac{1}{2}hA_x - \frac{1}{2}hA_y\right]U_{n+1} = \left[I + \frac{1}{2}hA_x + \frac{1}{2}hA_y\right]U_n + h(\vec{bc}) + O(h^3) .$$

From both sides subtract

$$\left[I - \frac{1}{2}hA_x - \frac{1}{2}hA_y\right]U_n$$

---

[9] A block tridiagonal solver is similar to a scalar solver except that small block matrix operations replace the scalar ones, and matrix multiplications do not commute.

leaving the equality unchanged. Then, using the standard definition of the difference operator Δ,

$$\Delta U_n = U_{n+1} - U_n$$

one finds

$$\left[I - \frac{1}{2}hA_x - \frac{1}{2}hA_y\right]\Delta U_n = h\left[A_{x+y}U_n + (\vec{bc})\right] + O(h^3) . \quad (12.28)$$

Notice that the right side of this equation is the product of h and a term that is identical to the right side of (12.13), our original ODE. Thus, if (12.28) converges, it is guaranteed to converge to the correct steady-state solution of the ODE. Now we can factor (12.28) and maintain second-order accuracy. We arrive at the expression

$$\left[I - \frac{1}{2}hA_x\right]\left[I - \frac{1}{2}hA_y\right]\Delta U_n = h\left[A_{x+y}U_n + (\vec{bc})\right] + O(h^3) . \quad (12.29)$$

This is the delta form of a factored, second-order, 2D equation.

The point at which the factoring is made may not affect the order of time-accuracy, but it can have a profound effect on the stability and convergence properties of a method. For example, the unfactored form of a first-order method derived from the implicit Euler time march is given by (12.20), and if it is immediately factored, the factored form is presented in (12.21). On the other hand, the *delta form* of the unfactored (12.20) is

$$[I - hA_x - hA_y]\Delta U_n = h\left[A_{x+y}U_n + (\vec{bc})\right]$$

and *its* factored form becomes[10]

$$[I - hA_x][I - hA_y]\Delta U_n = h\left[A_{x+y}U_n + (\vec{bc})\right] . \quad (12.30)$$

In spite of the similarities in derivation, we will see in the next chapter that the convergence properties of (12.21) and (12.30) are vastly different.

Exercises

12.1 Consider the 1D heat equation:

$$\frac{\partial u}{\partial t} = \nu \frac{\partial^2 u}{\partial x^2} , \qquad 0 \le x \le 9 .$$

Let $u(0,t) = 0$ and $u(9,t) = 0$, so that we can simplify the boundary conditions. Assume that second-order central differencing is used, i.e.

---

[10] Notice that the only difference between the second-order method given by (12.29) and the first-order method given by (12.30) is the appearance of the factor $1/2$ on the left side of the second-order method.

$$(\delta_{xx}u)_j = \frac{1}{\Delta x^2}(u_{j-1} - 2u_j + u_{j+1}) .$$

The uniform grid has $\Delta x = 1$ and 8 interior points.

(a) *Space vector definition*

    i. What is the space vector for the natural ordering (monotonically increasing in index), $u^{(1)}$? Only include the interior points.

    ii. If we reorder the points with the odd points first and then the even points, write the space vector, $u^{(2)}$.

    iii. Write down the permutation matrices, P_{12} and P_{21}.

    iv. The generic ODE representing the discrete form of the heat equation is

$$\frac{du^{(1)}}{dt} = A_1 u^{(1)} + f .$$

Write down the matrix A_1. (Note $f = 0$ due to the boundary conditions.) Next find the matrix A_2 such that

$$\frac{du^{(2)}}{dt} = A_2 u^{(2)} .$$

Note that A_2 can be written as

$$A_2 = \left[\begin{array}{c|c} D & U^{\mathrm{T}} \\ \hline U & D \end{array} \right] .$$

Define D and U.

    v. Applying implicit Euler time marching, write the delta form of the implicit algorithm. Comment on the form of the resulting implicit matrix operator.

(b) *System definition*

In part (a), we defined $u^{(1)}, u^{(2)}, A_1, A_2, P_{12}$, and P_{21} which partition the odd points from the even points. We can put such a partitioning to use. First define extraction operators

$$I^{(o)} = \left[\begin{array}{cccc|cccc} 1 & 0 & 0 & 0 & 0 & 0 & 0 & 0 \\ 0 & 1 & 0 & 0 & 0 & 0 & 0 & 0 \\ 0 & 0 & 1 & 0 & 0 & 0 & 0 & 0 \\ 0 & 0 & 0 & 1 & 0 & 0 & 0 & 0 \\ \hline 0 & 0 & 0 & 0 & 0 & 0 & 0 & 0 \\ 0 & 0 & 0 & 0 & 0 & 0 & 0 & 0 \\ 0 & 0 & 0 & 0 & 0 & 0 & 0 & 0 \\ 0 & 0 & 0 & 0 & 0 & 0 & 0 & 0 \end{array} \right] = \left[\begin{array}{c|c} I_4 & 0_4 \\ \hline 0_4 & 0_4 \end{array} \right] ,$$

$$
I^{(e)} = \begin{bmatrix}
0 & 0 & 0 & 0 & 0 & 0 & 0 & 0 \\
0 & 0 & 0 & 0 & 0 & 0 & 0 & 0 \\
0 & 0 & 0 & 0 & 0 & 0 & 0 & 0 \\
0 & 0 & 0 & 0 & 0 & 0 & 0 & 0 \\
\hline
0 & 0 & 0 & 0 & 1 & 0 & 0 & 0 \\
0 & 0 & 0 & 0 & 0 & 1 & 0 & 0 \\
0 & 0 & 0 & 0 & 0 & 0 & 1 & 0 \\
0 & 0 & 0 & 0 & 0 & 0 & 0 & 1
\end{bmatrix}
= \left[\begin{array}{c|c}
0_4 & 0_4 \\
\hline
0_4 & I_4
\end{array} \right] ,
$$

which extract the odd and even points from $u^{(2)}$ as follows: $u^{(o)} = I^{(o)} u^{(2)}$ and $u^{(e)} = I^{(e)} u^{(2)}$.

   i. Beginning with the ODE written in terms of $u^{(2)}$, define a splitting $A_2 = A_o + A_e$, such that A_o operates only on the odd terms, and A_e operates only on the even terms. Write out the matrices A_o and A_e. Also, write them in terms of D and U defined above.

   ii. Apply implicit Euler time marching to the split ODE. Write down the delta form of the algorithm and the factored delta form. Comment on the order of the error terms.

   iii. Examine the implicit operators for the factored delta form. Comment on their form. You should be able to argue that these are now triangular matrices (a lower and an upper one). Comment on the solution process this gives us relative to the direct inversion of the original system.

13. Analysis
of Split and Factored Forms

In Section 4.4 we introduced the concept of the representative equation and used it in Chapter 7 to study the stability, accuracy, and convergence properties of time-marching schemes. The question is: Can we find a similar equation that will allow us to evaluate the stability and convergence properties of split and factored schemes? The answer is yes – for certain forms of linear model equations.

The analysis in this chapter is useful for estimating the stability and steady-state properties of a wide variety of time-marching schemes that are variously referred to as time-split, fractional-step, hybrid, and (approximately) factored. When these methods are applied to practical problems, the results found from this analysis are neither necessary nor sufficient to guarantee stability. However, if the results indicate that a method has an instability, the method is probably not suitable for practical use.

13.1 The Representative Equation
for Circulant Operators

Consider linear PDE's with coefficients that are fixed in both space and time and with boundary conditions that are periodic. We have seen that under these conditions a semi-discrete approach can lead to circulant matrix difference operators, and we discussed circulant eigensystems[1] in Section 4.3. In this and the following section we assume circulant systems, and our analysis *depends critically on the fact that all circulant matrices commute and have a common set of eigenvectors.*

Suppose, as a result of space differencing the PDE, we arrive at a set of ODE's that can be written

$$\frac{d\vec{u}}{dt} = A_a \vec{u} + A_b \vec{u} - \vec{f}(t) \,, \tag{13.1}$$

where A_a and A_b are circulant matrices. Since both matrices have the same set of eigenvectors, we can use the arguments made in Section 4.2.3 to uncouple the set and form the set of M independent equations

---

[1] See also the discussion on Fourier stability analysis in Section 7.7.

$$w_1' = (\lambda_a + \lambda_b)_1 w_1 - g_1(t)$$

$$\vdots$$

$$w_m' = (\lambda_a + \lambda_b)_m w_m - g_m(t) \tag{13.2}$$

$$\vdots$$

$$w_M' = (\lambda_a + \lambda_b)_M w_M - g_M(t) \,.$$

The solution of the mth line is

$$w_m(t) = c_m e^{(\lambda_a + \lambda_b)_m t} + P.S. \,.$$

Note that each λ_a pairs with one, and only one,[2] λ_b, since they must share a common eigenvector. This suggests (see Section 4.4):

The representative equation for split, circulant systems is

$$\frac{du}{dt} = [\lambda_a + \lambda_b + \lambda_c + \cdots]u + ae^{\mu t} \,, \tag{13.3}$$

where $\lambda_a + \lambda_b + \lambda_c + \cdots$ is the sum of the eigenvalues in A_a , A_b , A_c , $\cdots$ that share the same eigenvector.

13.2 Example Analysis of Circulant Systems

13.2.1 Stability Comparisons of Time-Split Methods

Consider as an example the linear convection-diffusion equation

$$\frac{\partial u}{\partial t} + a\frac{\partial u}{\partial x} = \nu\frac{\partial^2 u}{\partial x^2} \,. \tag{13.4}$$

If the space differencing takes the form

$$\frac{d\vec{u}}{dt} = -\frac{a}{2\Delta x}B_p(-1,0,1)\vec{u} + \frac{\nu}{\Delta x^2}B_p(1,-2,1)\vec{u} \tag{13.5}$$

the convection matrix operator and the diffusion matrix operator, can be represented by the eigenvalues λ_c and λ_d, respectively, where (see Section 4.3.2)

$$(\lambda_c)_m = \frac{ia}{\Delta x}\sin\theta_m$$

$$(\lambda_d)_m = -\frac{4\nu}{\Delta x^2}\sin^2\frac{\theta_m}{2} \,. \tag{13.6}$$

---

[2] This is to be contrasted to the developments found later in the analysis of 2D equations.

In these equations, $\theta_m = 2m\pi/M$, and $m = 0, 1, \cdots, M - 1$, so that $0 \leq \theta_m \leq 2\pi$. Using these values and the representative equation (13.4), we can analyze the stability of the two forms of simple time-splitting discussed in Section 12.2. In this section we refer to these as

1. the explicit-implicit Euler method, (12.10),
2. the explicit-explicit Euler method, (12.8).

(1) The Explicit-Implicit Method. When applied to (13.4), the characteristic polynomial of this method is

$$P(E) = (1 - h\lambda_d)E - (1 + h\lambda_c) .$$

This leads to the principal σ-root

$$\sigma = \frac{1 + i\dfrac{ah}{\Delta x}\sin\theta_m}{1 + 4\dfrac{h\nu}{\Delta x^2}\sin^2\dfrac{\theta_m}{2}} ,$$

where we have made use of (13.6) to quantify the eigenvalues. Now introduce the dimensionless numbers

$$C_n = \frac{ah}{\Delta x}, \text{ Courant number}$$

$$R_\Delta = \frac{a\Delta x}{\nu}, \text{ mesh Reynolds number}$$

and we can write for the absolute value of σ

$$|\sigma| = \frac{\sqrt{1 + C_n^2 \sin^2 \theta_m}}{1 + 4\dfrac{C_n}{R_\Delta}\sin^2\dfrac{\theta_m}{2}} , \quad 0 \leq \theta_m \leq 2\pi . \tag{13.7}$$

A simple numerical parametric study of (13.7) shows that the critical range of θ_m for any combination of C_n and R_Δ occurs when θ_m is near 0 (or 2π). From this we find that the condition on C_n and R_Δ that makes $|\sigma| \approx 1$ is

$$[1 + C_n^2 \sin^2 \epsilon] = \left[1 + 4\frac{C_n}{R_\Delta}\sin^2\frac{\epsilon}{2}\right]^2 .$$

As $\epsilon \to 0$ this gives the stability region

$$C_n < \frac{2}{R_\Delta}$$

which is bounded by a hyperbola and shown in Figure 13.1.

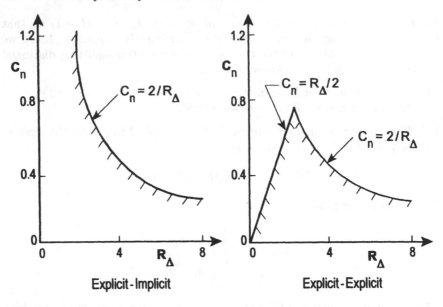

Fig. 13.1. Stability regions for two simple time-split methods

(2) The Explicit-Explicit Method. An analysis similar to the one given above shows that this method produces

$$|\sigma| = \sqrt{1 + C_n^2 \sin^2\theta_m \left[1 - 4\frac{C_n}{R_\Delta}\sin^2\frac{\theta_m}{2}\right]} \quad , \quad 0 \le \theta_m \le 2\pi \ .$$

Again a simple numerical parametric study shows that this has two critical ranges of θ_m, one near 0, which yields the same result as in the previous example, and the other near π, which produces the constraint that

$$C_n < \frac{1}{2}R_\Delta \quad \text{for} \quad R_\Delta \le 2 \ .$$

The resulting stability boundary is also shown in Figure 13.1. The totally explicit factored method has a much smaller region of stability when R_Δ is small, as we should have expected.

13.2.2 Analysis of a Second-Order Time-Split Method

Next let us analyze a more practical method that has been used in serious computational analysis of turbulent flows. This method applies to a flow in which there is a combination of diffusion and periodic convection. The convection term is treated explicitly using the second-order Adams–Bashforth method. The diffusion term is integrated implicitly using the trapezoidal method. Our model equation is again the linear convection-diffusion equation (13.4), which we split in the fashion of (13.5). In order to evaluate the accuracy, as well as the stability, we include the forcing function in the representative equation and study the effect of our hybrid, time-marching method on the equation

$$u' = \lambda_c u + \lambda_d u + a e^{\mu t} \ .$$

First let us find expressions for the two polynomials, $P(E)$ and $Q(E)$. The characteristic polynomial follows from the application of the method to the homogeneous equation, thus

$$u_{n+1} = u_n + \frac{1}{2} h \lambda_c (3u_n - u_{n-1}) + \frac{1}{2} h \lambda_d (u_{n+1} + u_n) \ .$$

This produces

$$P(E) = \left(1 - \frac{1}{2} h \lambda_d\right) E^2 - \left(1 + \frac{3}{2} h \lambda_c + \frac{1}{2} h \lambda_d\right) E + \frac{1}{2} h \lambda_c \ .$$

The form of the particular polynomial depends upon whether the forcing function is carried by the AB2 method or by the trapezoidal method. In the former case it is

$$Q(E) = \frac{1}{2} h (3E - 1) \tag{13.8}$$

and in the latter

$$Q(E) = \frac{1}{2} h (E^2 + E) \ . \tag{13.9}$$

Accuracy. From the characteristic polynomial we see that there are two σ-roots, and they are given by the equation

$$\sigma = \frac{1 + \frac{3}{2} h \lambda_c + \frac{1}{2} h \lambda_d \pm \sqrt{\left(1 + \frac{3}{2} h \lambda_c + \frac{1}{2} h \lambda_d\right)^2 - 2 h \lambda_c \left(1 - \frac{1}{2} h \lambda_d\right)}}{2 \left(1 - \frac{1}{2} h \lambda_d\right)} \ . \tag{13.10}$$

The principal σ-root follows from the plus sign, and one can show that

$$\sigma_1 = 1 + (\lambda_c + \lambda_d) h + \frac{1}{2} (\lambda_c + \lambda_d)^2 h^2 + \frac{1}{4} (\lambda_d^3 + \lambda_c \lambda_d^2 - \lambda_c^2 \lambda_d - \lambda_c^3) h^3 \ .$$

From this equation it is clear that $\frac{1}{6} \lambda^3 = \frac{1}{6} (\lambda_c + \lambda_d)^3$ does not match the coefficient of h^3 in σ_1, so

$$er_\lambda = O(h^3) \ .$$

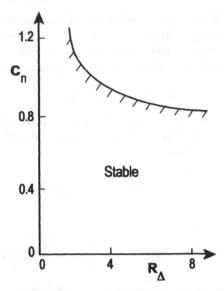

Fig. 13.2. Stability region for the second-order time-split method

Using $P(e^{\mu h})$ and $Q(e^{\mu h})$ to evaluate er_μ in Section 6.6.3, one can show that

$$er_\mu = O(h^3)$$

using *either* (13.8) *or* (13.9). These results show that, for the model equation, the hybrid method retains the second-order accuracy of its individual components.

Stability. The stability of the method can be found from (13.10) by a parametric study of C_n and R_Δ defined above (13.7). This was carried out in a manner similar to that used to find the stability boundary of the first-order explicit-implicit method in Section 13.2.1. The results are plotted in Figure 13.2. For values of $R_\Delta \geq 2$ this second-order method has a much greater region of stability than the first-order explicit-implicit method given by (12.10) and shown in Figure 13.1.

13.3 The Representative Equation for Space-Split Operators

Consider the 2D model equations[3]

$$\frac{\partial u}{\partial t} = \frac{\partial^2 u}{\partial x^2} + \frac{\partial^2 u}{\partial y^2} \tag{13.11}$$

---

[3] The extension of the following to 3D is simple and straightforward.

and

$$\frac{\partial u}{\partial t} + a_x \frac{\partial u}{\partial x} + a_y \frac{\partial u}{\partial y} = 0 . \tag{13.12}$$

Reduce either of these, by means of spatial differencing approximations, to the coupled set of ODE's:

$$\frac{dU}{dt} = [A_x + A_y]U + (\vec{bc}) \tag{13.13}$$

for the space vector U. The form of the A_x and A_y matrices for three-point central differencing schemes are shown in (12.18) and (12.19) for the 3×4 mesh shown in Section 12.3.1. Let us inspect the structure of these matrices closely to see how we can diagonalize $[A_x + A_y]$ in terms of the individual eigenvalues of the two matrices considered separately.

First we write these matrices in the form

$$A_x^{(x)} = \begin{bmatrix} B & & \\ & B & \\ & & B \end{bmatrix} \quad A_y^{(x)} = \begin{bmatrix} \tilde{b}_0 \cdot I & \tilde{b}_1 \cdot I & \\ \tilde{b}_{-1} \cdot I & \tilde{b}_0 \cdot I & \tilde{b}_1 \cdot I \\ & \tilde{b}_{-1} \cdot I & \tilde{b}_0 \cdot I \end{bmatrix} ,$$

where B is a banded matrix of the form $B(b_{-1}, b_0, b_1)$. Now find the block eigenvector matrix that diagonalizes B and use it to diagonalize $A_x^{(x)}$. Thus

$$\begin{bmatrix} X^{-1} & & \\ & X^{-1} & \\ & & X^{-1} \end{bmatrix} \begin{bmatrix} B & & \\ & B & \\ & & B \end{bmatrix} \begin{bmatrix} X & & \\ & X & \\ & & X \end{bmatrix} = \begin{bmatrix} \Lambda & & \\ & \Lambda & \\ & & \Lambda \end{bmatrix} ,$$

where

$$\Lambda = \begin{bmatrix} \lambda_1 & & & \\ & \lambda_2 & & \\ & & \lambda_3 & \\ & & & \lambda_4 \end{bmatrix} .$$

Notice that the matrix $A_y^{(x)}$ is *transparent* to this transformation. That is, if we set $\mathbf{X} \equiv \mathrm{diag}(X)$

$$\mathbf{X}^{-1} \begin{bmatrix} \tilde{b}_0 \cdot I & \tilde{b}_1 \cdot I & \\ \tilde{b}_{-1} \cdot I & \tilde{b}_0 \cdot I & \tilde{b}_1 \cdot I \\ & \tilde{b}_{-1} \cdot I & \tilde{b}_0 \cdot I \end{bmatrix} \mathbf{X} = \begin{bmatrix} \tilde{b}_0 \cdot I & \tilde{b}_1 \cdot I & \\ \tilde{b}_{-1} \cdot I & \tilde{b}_0 \cdot I & \tilde{b}_1 \cdot I \\ & \tilde{b}_{-1} \cdot I & \tilde{b}_0 \cdot I \end{bmatrix} .$$

One now permutes the transformed system to the y-vector data-base using the permutation matrix defined by (12.12). There results

$$P_{yx} \cdot \mathbf{X}^{-1} \left[A_x^{(x)} + A_y^{(x)} \right] \mathbf{X} \cdot P_{xy}$$

$$= \begin{bmatrix} \lambda_1 \cdot I & & & \\ & \lambda_2 \cdot I & & \\ & & \lambda_3 \cdot I & \\ & & & \lambda_4 \cdot I \end{bmatrix} + \begin{bmatrix} \tilde{B} & & & \\ & \tilde{B} & & \\ & & \tilde{B} & \\ & & & \tilde{B} \end{bmatrix} , \tag{13.14}$$

where $\tilde{B}$ is the banded tridiagonal matrix $B(\tilde{b}_{-1}, \tilde{b}_0, \tilde{b}_1)$ (see the bottom of (12.19)). Next find the eigenvectors $\tilde{X}$ that diagonalize the $\tilde{B}$ blocks. Let $\tilde{\mathbf{B}} \equiv \text{diag}(\tilde{B})$ and $\tilde{\mathbf{X}} \equiv \text{diag}(\tilde{X})$ and form the second transformation

$$\tilde{\mathbf{X}}^{-1}\tilde{\mathbf{B}}\tilde{\mathbf{X}} = \begin{bmatrix} \tilde{\Lambda} & & \\ & \tilde{\Lambda} & \\ & & \tilde{\Lambda} \end{bmatrix} \quad , \quad \tilde{\Lambda} = \begin{bmatrix} \tilde{\lambda}_1 & & \\ & \tilde{\lambda}_2 & \\ & & \tilde{\lambda}_3 \end{bmatrix} .$$

This time, by the same argument as before, the first matrix on the right side of (13.14) is transparent to the transformation, so the final result is the complete diagonalization of the matrix A_{x+y}:

$$\left[\tilde{\mathbf{X}}^{-1} \cdot P_{yx} \cdot \mathbf{X}^{-1} \right] \left[A^{(x)}_{x+y} \right] \left[\mathbf{X} \cdot P_{xy} \cdot \tilde{\mathbf{X}} \right]$$

$$+ \begin{bmatrix} \lambda_1 I + \tilde{\Lambda} & & & \\ & \lambda_2 I + \tilde{\Lambda} & & \\ & & \lambda_3 I + \tilde{\Lambda} & \\ & & & \lambda_4 I + \tilde{\Lambda} \end{bmatrix} . \tag{13.15}$$

It is important to notice that:

- The diagonal matrix on the right side of (13.15) contains every possible combination of the individual eigenvalues of B and $\tilde{B}$.

Now we are ready to present the representative equation for 2D systems. First reduce the PDE to ODE's by some choice[4] of space differencing. This results in a spatially split A matrix formed from the subsets

$$A^{(x)}_x = \text{diag}(B) \quad , \quad A^{(y)}_y = \text{diag}(\tilde{B}) , \tag{13.16}$$

where B and $\tilde{B}$ are any two matrices that have linearly independent eigenvectors (this puts some constraints on the choice of differencing schemes).

Although A_x and A_y do commute, this fact, by itself, does not ensure the property of "all possible combinations". To obtain the latter property the structure of the matrices is important. The block matrices B and $\tilde{B}$ can be either circulant or noncirculant; in both cases we are led to the final result:

---

[4] We have used 3-point central differencing in our example, but this choice was for convenience only, and its use is not necessary to arrive at (13.15.)

The 2D representative equation for model linear systems is

$$\frac{du}{dt} = [\lambda_x + \lambda_y]u + ae^{\mu t} \; ,$$

where λ_x and λ_y are *any combination* of eigenvalues from A_x and A_y, a and μ are (possibly complex) constants, and where A_x and A_y satisfy the conditions in (13.16).

Often we are interested in finding the value of, and the convergence rate to, the steady-state solution of the representative equation. In that case we set $\mu = 0$ and use the simpler form

$$\frac{du}{dt} = [\lambda_x + \lambda_y]u + a \; , \tag{13.17}$$

which has the exact solution

$$u(t) = ce^{(\lambda_x + \lambda_y)t} - \frac{a}{\lambda_x + \lambda_y} \; . \tag{13.18}$$

13.4 Example Analysis of the 2D Model Equation

In the following we analyze four different methods for finding a fixed, steady-state solution to the 2D representative equation (13.17). In each case we examine

(1) the stability,
(2) the accuracy of the fixed, steady-state solution,
(3) the convergence rate to reach the steady state.

13.4.1 The Unfactored Implicit Euler Method

Consider first this unfactored, first-order scheme, which can then be used as a reference case for comparison with the various factored ones. The form of the method is given by (12.20), and when it is applied to the representative equation, we find

$$(1 - h\lambda_x - h\lambda_y)u_{n+1} = u_n + ha \; ,$$

from which

$$P(E) = (1 - h\lambda_x - h\lambda_y)E - 1$$
$$Q(E) = h \tag{13.19}$$

giving the solution

$$u_n = c\left(\frac{1}{1 - h\,\lambda_x - h\,\lambda_y}\right)^n - \frac{a}{\lambda_x + \lambda_y}\ .$$

Like its counterpart in the 1D case, this method

(1) is unconditionally stable,
(2) produces the exact (see (13.18)) steady-state solution (of the ODE) for any h,
(3) converges very rapidly to the steady state when h is large.

Unfortunately, however, use of this method for 2D problems is generally impractical for reasons discussed in Section 12.5.

13.4.2 The Factored Nondelta Form of the Implicit Euler Method

Now apply the factored Euler method given by (12.21) to the 2D representative equation. There results

$$(1 - h\,\lambda_x)(1 - h\,\lambda_y)u_{n+1} = u_n + ha\ ,$$

from which

$$\begin{aligned} P(E) &= (1 - h\,\lambda_x)(1 - h\,\lambda_y)E - 1 \\ Q(E) &= h \end{aligned} \tag{13.20}$$

giving the solution

$$u_n = c\left[\frac{1}{(1 - h\,\lambda_x)(1 - h\,\lambda_y)}\right]^n - \frac{a}{\lambda_x + \lambda_y - h\lambda_x\lambda_y}\ .$$

We see that this method

(1) is unconditionally stable,
(2) produces a steady-state solution that depends on the choice of h,
(3) converges rapidly to a steady state for large h, but the converged solution is completely wrong.

The method requires far less storage then the unfactored form. However, it is not very useful since its transient solution is only first-order accurate and, if one tries to take advantage of its rapid convergence rate, the converged value is meaningless.

13.4.3 The Factored Delta Form of the Implicit Euler Method

Next apply (12.30) to the 2D representative equation. One finds

$$(1 - h\lambda_x)(1 - h\lambda_y)(u_{n+1} - u_n) = h(\lambda_x u_n + \lambda_y u_n + a) ,$$

which reduces to

$$(1 - h\lambda_x)(1 - h\lambda_y)u_{n+1} = (1 + h^2\lambda_x\lambda_y)u_n + ha ,$$

and this has the solution

$$u_n = c\left[\frac{1 + h^2\lambda_x\lambda_y}{(1 - h\lambda_x)(1 - h\lambda_y)}\right]^n - \frac{a}{\lambda_x + \lambda_y} .$$

This method

(1) is unconditionally stable,
(2) produces the exact steady-state solution for any choice of h,
(3) converges very slowly to the steady-state solution for large values of h,
   since $|\sigma| \to 1$ as $h \to \infty$.

Like the factored nondelta form, this method demands far less storage than the unfactored form, as discussed in Section 12.5. The correct steady solution is obtained, but convergence is not nearly as rapid as that of the unfactored form.

13.4.4 The Factored Delta Form of the Trapezoidal Method

Finally consider the delta form of a second-order time-accurate method. Applying (12.29) to the representative equation, one finds

$$\left(1 - \frac{1}{2}h\lambda_x\right)\left(1 - \frac{1}{2}h\lambda_y\right)(u_{n+1} - u_n) = h(\lambda_x u_n + \lambda_y u_n + a) ,$$

which reduces to

$$\left(1 - \frac{1}{2}h\lambda_x\right)\left(1 - \frac{1}{2}h\lambda_y\right)u_{n+1} = \left(1 + \frac{1}{2}h\lambda_x\right)\left(1 + \frac{1}{2}h\lambda_y\right)u_n + ha ,$$

and this has the solution

$$u_n = c\left[\frac{\left(1 + \frac{1}{2}h\lambda_x\right)\left(1 + \frac{1}{2}h\lambda_y\right)}{\left(1 - \frac{1}{2}h\lambda_x\right)\left(1 - \frac{1}{2}h\lambda_y\right)}\right]^n - \frac{a}{\lambda_x + \lambda_y} .$$

This method

(1) is unconditionally stable,
(2) produces the exact steady-state solution for any choice of h,
(3) converges very slowly to the steady–state solution for large values of h, since $|\sigma| \to 1$ as $h \to \infty$.

All of these properties are identical to those found for the factored delta form of the implicit Euler method. Since it is second-order in time, it can be used when time accuracy is desired, and the factored delta form of the implicit Euler method can be used when a converged steady state is all that is required.[5] A brief inspection of (12.25) and (12.26) should be enough to convince the reader that the σ's produced by those methods are identical to the σ produced by this method.

13.5 Example Analysis of the 3D Model Equation

The arguments in Section 13.3 generalize to three dimensions and, under the conditions given in (13.16) with an $A_z^{(z)}$ included, the model 3D cases[6] have the following representative equation (with $\mu = 0$):

$$\frac{du}{dt} = [\lambda_x + \lambda_y + \lambda_z]u + a . \tag{13.21}$$

Let us analyze a 2nd-order accurate, factored delta form using this equation. First apply the trapezoidal method:

$$u_{n+1} = u_n + \frac{1}{2}h[(\lambda_x + \lambda_y + \lambda_z)u_{n+1} + (\lambda_x + \lambda_y + \lambda_z)u_n + 2a] .$$

Rearrange terms:

$$\left[1 - \frac{1}{2}h(\lambda_x + \lambda_y + \lambda_z)\right]u_{n+1} = \left[1 + \frac{1}{2}h(\lambda_x + \lambda_y + \lambda_z)\right]u_n + ha .$$

Put this in delta form:

$$\left[1 - \frac{1}{2}h(\lambda_x + \lambda_y + \lambda_z)\right]\Delta u_n = h[(\lambda_x + \lambda_y + \lambda_z)u_n + a] .$$

Now factor the left side:

$$\left(1 - \frac{1}{2}h\lambda_x\right)\left(1 - \frac{1}{2}h\lambda_y\right)\left(1 - \frac{1}{2}h\lambda_z\right)\Delta u_n = h\left[(\lambda_x + \lambda_y + \lambda_z)u_n + a\right] .$$

$$\tag{13.22}$$

---

[5] In practical codes, the value of h on the left side of the implicit equation is literally switched from h to $\frac{1}{2}h$.

[6] Equations (13.11) and (13.12), each with an additional term.

This preserves second order accuracy, since the error terms

$$-\frac{1}{4}h^2(\lambda_x\lambda_y + \lambda_x\lambda_z + \lambda_y\lambda_z)\Delta u_n \quad \text{and} \quad \frac{1}{8}h^3\lambda_x\lambda_y\lambda_z$$

are both $O(h^3)$. One can derive the characteristic polynomial for (13.22), find the σ-root, and write the solution either in the form

$$u_n = c\left[\frac{1 + \frac{1}{2}h(\lambda_x + \lambda_y + \lambda_z) + \frac{1}{4}h^2(\lambda_x\lambda_y + \lambda_x\lambda_z + \lambda_y\lambda_z) - \frac{1}{8}h^3\lambda_x\lambda_y\lambda_z}{1 - \frac{1}{2}h(\lambda_x + \lambda_y + \lambda_z) + \frac{1}{4}h^2(\lambda_x\lambda_y + \lambda_x\lambda_z + \lambda_y\lambda_z) - \frac{1}{8}h^3\lambda_x\lambda_y\lambda_z}\right]^n$$
$$-\frac{a}{\lambda_x + \lambda_y + \lambda_z} \tag{13.23}$$

or in the form

$$u_n = c\left[\frac{\left(1 + \frac{1}{2}h\lambda_x\right)\left(1 + \frac{1}{2}h\lambda_y\right)\left(1 + \frac{1}{2}h\lambda_z\right) - \frac{1}{4}h^3\lambda_x\lambda_y\lambda_z}{\left(1 - \frac{1}{2}h\lambda_x\right)\left(1 - \frac{1}{2}h\lambda_y\right)\left(1 - \frac{1}{2}h\lambda_z\right)}\right]^n$$
$$-\frac{a}{\lambda_x + \lambda_y + \lambda_z} \cdot \tag{13.24}$$

It is interesting to notice that a Taylor series expansion of (13.24) results in

$$\sigma = 1 + h(\lambda_x + \lambda_y + \lambda_z) + \frac{1}{2}h^2(\lambda_x + \lambda_y + \lambda_z)^2$$
$$+\frac{1}{4}h^3\left[\lambda_z^3 + (2\lambda_y + 2\lambda_x) + (2\lambda_y^2 + 3\lambda_x\lambda_y + 2\lambda_y^2)\right.$$
$$\left. + \lambda_y^3 + 2\lambda_x\lambda_y^2 + 2\lambda_x^2\lambda_y + \lambda_x^3\right] + \cdots , \tag{13.25}$$

which verifies the second-order accuracy of the factored form. Furthermore, clearly, if the method converges, it converges to the proper steady state.[7]

With regards to stability, it follows from (13.23) that, if all the λ's are real and negative, the method is stable for all h. This makes the method *unconditionally stable* for the 3D *diffusion* model when it is centrally differenced in space.

Now consider what happens when we apply this method to the biconvection model, the 3D form of (13.12) with periodic boundary conditions. In this case, central differencing causes all of the λ's to be imaginary with spectrums that include both positive and negative values. *Remember that in our analysis we must consider every possible combination of these eigenvalues.* First write the σ-root in (13.23) in the form

$$\sigma = \frac{1 + i\alpha - \beta + i\gamma}{1 - i\alpha - \beta + i\gamma},$$

---

[7] However, we already knew this because we chose the delta form.

where α, β and γ are real numbers that can have any sign. Now we can always find one combination of the λ's for which α, and γ are both positive. In that case since the absolute value of the product is the product of the absolute values

$$|\sigma|^2 = \frac{(1-\beta)^2 + (\alpha+\gamma)^2}{(1-\beta)^2 + (\alpha-\gamma)^2} > 1$$

and the method is *unconditionally unstable* for the model *convection* problem.

From the above analysis one would come to the conclusion that the method represented by (13.22) should not be used for the 3D Euler equations. In practical cases, however, some form of dissipation is almost always added to methods that are used to solve the Euler equations, and our experience to date is that, in the presence of this dissipation, the instability disclosed above is too weak to cause trouble.

Exercises

13.1 Starting with the generic ODE,

$$\frac{du}{dt} = Au + f ,$$

we can split A as follows: $A = A_1 + A_2 + A_3 + A_4$. Applying implicit Euler time marching gives

$$\frac{u_{n+1} - u_n}{h} = A_1 u_{n+1} + A_2 u_{n+1} + A_3 u_{n+1} + A_4 u_{n+1} + f .$$

(a) Write the factored delta form. What is the error term?
(b) Instead of making all of the split terms implicit, leave two explicit:

$$\frac{u_{n+1} - u_n}{h} = A_1 u_{n+1} + A_2 u_n + A_3 u_{n+1} + A_4 u_n + f .$$

Write the resulting factored delta form and define the error terms.
(c) The scalar representative equation is

$$\frac{du}{dt} = (\lambda_1 + \lambda_2 + \lambda_3 + \lambda_4)u + a .$$

For the fully implicit scheme of Exercise 13.1(a), find the exact solution to the resulting scalar difference equation and comment on the stability, convergence, and accuracy of the converged steady-state solution.
(d) Repeat Exercise 13.1(c) for the explicit-implicit scheme of Exercise 13.1(b).

A. Useful Relations from Linear Algebra

A basic understanding of the fundamentals of linear algebra is crucial to our development of numerical methods, and it is assumed that the reader is at least familar with this subject area. Given below is some notation and some of the important relations.

A.1 Notation

(1) In the present context a vector is a vertical column or string. Thus

$$\vec{v} = \begin{bmatrix} v_1 \\ v_2 \\ \vdots \\ v_m \end{bmatrix}$$

and its transpose $\vec{v}^{\mathrm{T}}$ is the horizontal row

$$\vec{v}^{\mathrm{T}} = [v_1, v_2, v_3, \ldots, v_m] \quad , \quad \vec{v} = [v_1, v_2, v_3, \ldots, v_m]^{\mathrm{T}} .$$

(2) A general $m \times m$ matrix A can be written

$$A = (a_{ij}) = \begin{bmatrix} a_{11} & a_{12} & \cdots & a_{1m} \\ a_{21} & a_{22} & \cdots & a_{2m} \\ & & \ddots & \\ a_{m1} & a_{m2} & \cdots & a_{mm} \end{bmatrix} .$$

(3) An alternative notation for A is

$$A = [\vec{a}_1, \vec{a}_2, \ldots, \vec{a}_m]$$

and its transpose A^{T} is

$$A^{\mathrm{T}} = \begin{bmatrix} \vec{a}_1^{\mathrm{T}} \\ \vec{a}_2^{\mathrm{T}} \\ \vdots \\ \vec{a}_m^{\mathrm{T}} \end{bmatrix} .$$

(4) The inverse of a matrix (if it exists) is written A^{-1} and has the property that $A^{-1}A = AA^{-1} = I$, where I is the identity matrix.

A.2 Definitions

(1) A is *symmetric* if $A^{\mathrm{T}} = A$.

(2) A is *skew-symmetric* or *antisymmetric* if $A^{\mathrm{T}} = -A$.

(3) A is *diagonally dominant* if $a_{ii} \geq \sum_{j \neq i} |a_{ij}|$, $i = 1, 2, \ldots, m$, and $a_{ii} > \sum_{j \neq i} |a_{ij}|$ for at least one i.

(4) A is *orthogonal* if a_{ij} are real, and $A^{\mathrm{T}} A = A A^{\mathrm{T}} = I$.

(5) $\bar{A}$ is the *complex conjugate* of A.

(6) P is a *permutation* matrix if $P\vec{v}$ is a simple reordering of $\vec{v}$.

(7) The *trace* of a matrix is $\sum_i a_{ii}$.

(8) A is *normal* if $A^{\mathrm{T}} A = A A^{\mathrm{T}}$.

(9) $\det[A]$ is the determinant of A.

(10) A^{H} is the conjugate (or Hermitian) transpose of A.

(11) If

$$A = \begin{bmatrix} a & b \\ c & d \end{bmatrix}$$

then

$$\det[A] = ad - bc$$

and

$$A^{-1} = \frac{1}{\det[A]} \begin{bmatrix} d & -b \\ -c & a \end{bmatrix} .$$

A.3 Algebra

We consider only square matrices of the same dimension.

(1) A and B are equal if $a_{ij} = b_{ij}$ for all $i, j = 1, 2, \ldots, m$.

(2) $A + (B + C) = (C + A) + B$, etc.

(3) $sA = (sa_{ij})$ where s is a scalar.

(4) In general $AB \neq BA$.

(5) Transpose equalities:

$$(A + B)^{\mathrm{T}} = A^{\mathrm{T}} + B^{\mathrm{T}}$$
$$(A^{\mathrm{T}})^{\mathrm{T}} = A$$
$$(AB)^{\mathrm{T}} = B^{\mathrm{T}} A^{\mathrm{T}}$$

(6) Inverse equalities (if the inverse exists):

$$(A^{-1})^{-1} = A$$
$$(AB)^{-1} = B^{-1} A^{-1}$$
$$(A^{\mathrm{T}})^{-1} = (A^{-1})^{\mathrm{T}}$$

(7) Any matrix A can be expressed as the sum of a symmetric and a skew-symmetric matrix. Thus

$$A = \frac{1}{2}(A + A^{\mathrm{T}}) + \frac{1}{2}(A - A^{\mathrm{T}}) .$$

A.4 Eigensystems

(1) The eigenvalue problem for a matrix A is defined as

$$A\vec{x} = \lambda\vec{x} \quad \text{or} \quad [A - \lambda I]\vec{x} = 0$$

and the generalized eigenvalue problem, including the matrix B, as

$$A\vec{x} = \lambda B\vec{x} \quad \text{or} \quad [A - \lambda B]\vec{x} = 0 \ .$$

(2) If a square matrix with real elements is symmetric, its eigenvalues are all real. If it is skew-symmetric, they are all imaginary.

(3) Gershgorin's theorem: The eigenvalues of a matrix lie in the complex plane in the union of circles having centers located by the diagonals with radii equal to the sum of the absolute values of the corresponding off-diagonal row elements.

(4) In general, an $m \times m$ matrix A has $n_{\vec{x}}$ linearly independent eigenvectors with $n_{\vec{x}} \le m$ and n_{λ} distinct eigenvalues (λ_i) with $n_{\lambda} \le n_{\vec{x}} \le m$.

(5) A set of eigenvectors is said to be linearly independent if

$$a \cdot \vec{x}_m + b \cdot \vec{x}_n \neq \vec{x}_k \ , \quad m \neq n \neq k$$

for any complex a and b and for all combinations of vectors in the set.

(6) If A posseses m linearly independent eigenvectors then A is diagonalizable, i.e.

$$X^{-1}AX = \Lambda \ ,$$

where X is a matrix whose columns are the eigenvectors,

$$X = [\vec{x}_1, \vec{x}_2, \ldots, \vec{x}_m] \ ,$$

and Λ is the diagonal matrix

$$\Lambda = \begin{bmatrix} \lambda_1 & 0 & \cdots & 0 \\ 0 & \lambda_2 & \ddots & \vdots \\ \vdots & \ddots & \ddots & 0 \\ 0 & \cdots & 0 & \lambda_m \end{bmatrix} \ .$$

If A can be diagonalized, its eigenvectors completely span the space, and A is said to have a *complete* eigensystem.

(7) If A has m distinct eigenvalues, then A is always diagonalizable. With each distinct eigenvalue there is one associated eigenvector, and this eigenvector cannot be formed from a linear combination of any of the other eigenvectors.

(8) In general, the eigenvalues of a matrix may not be distinct, in which case the possibility exists that it cannot be diagonalized. If the eigenvalues of a matrix are not distinct, but all of the eigenvectors are linearly independent, the matrix is said to be *derogatory*, and it can still be diagonalized.

(9) If a matrix does not have a complete set of linearly independent eigenvectors, it cannot be diagonalized. The eigenvectors of such a matrix cannot span the space, and the matrix is said to have a *defective* eigensystem.

(10) Defective matrices cannot be diagonalized, but they can still be put into a compact form by a similarity transform, S, such that

$$
J = S^{-1}AS = \begin{bmatrix} J_1 & 0 & \cdots & 0 \\ 0 & J_2 & \ddots & \vdots \\ \vdots & \ddots & \ddots & 0 \\ 0 & \cdots & 0 & J_k \end{bmatrix},
$$

where there are k linearly independent eigenvectors, and J_i is either a Jordan subblock or λ_i.

(11) A Jordan submatrix has the form

$$
J_i = \begin{bmatrix} \lambda_i & 1 & 0 & \cdots & 0 \\ 0 & \lambda_i & 1 & \ddots & \vdots \\ 0 & 0 & \lambda_i & \ddots & 0 \\ \vdots & & \ddots & \ddots & 1 \\ 0 & \cdots & 0 & 0 & \lambda_i \end{bmatrix}.
$$

(12) Use of the transform S is known as putting A into its *Jordan Canonical form*. A repeated root in a Jordan block is referred to as a *defective* eigenvalue. For each Jordan submatrix with an eigenvalue λ_i of multiplicity r, there exists one eigenvector. The other $r - 1$ vectors associated with this eigenvalue are referred to as *principal vectors*. The complete set of principal vectors and eigenvectors are all linearly independent.

(13) Note that if P is the permutation matrix

$$
P = \begin{bmatrix} 0 & 0 & 1 \\ 0 & 1 & 0 \\ 1 & 0 & 0 \end{bmatrix}, \qquad P^{\mathrm{T}} = P^{-1} = P
$$

then

$$
P^{-1} \begin{bmatrix} \lambda & 1 & 0 \\ 0 & \lambda & 1 \\ 0 & 0 & \lambda \end{bmatrix} P = \begin{bmatrix} \lambda & 0 & 0 \\ 1 & \lambda & 0 \\ 0 & 1 & \lambda \end{bmatrix}.
$$

(14) Some of the Jordan blocks may have the same eigenvalue. For example, the matrix

$$\left[\begin{array}{c}\begin{bmatrix}\lambda_1 & 1 & \\ & \lambda_1 & 1 \\ & & \lambda_1\end{bmatrix} & & & \\ & \lambda_1 & & \\ & & \begin{bmatrix}\lambda_1 & 1 \\ & \lambda_1\end{bmatrix} & \\ & & & \begin{bmatrix}\lambda_2 & 1 \\ & \lambda_2\end{bmatrix} \\ & & & & \lambda_3\end{array}\right]$$

is both defective and derogatory, having:

- nine eigenvalues,
- three distinct eigenvalues,
- three Jordan blocks,
- five linearly independent eigenvectors,
- three principal vectors with λ_1,
- one principal vector with λ_2.

A.5 Vector and Matrix Norms

(1) The *spectral radius* of a matrix A is symbolized by $\sigma(A)$ such that

$$\sigma(A) = |\sigma_m|_{\max} \, ,$$

where σ_m are the eigenvalues of the matrix A.

(2) A *p-norm* of the vector $\vec{v}$ is defined as

$$\|v\|_p = \left(\sum_{j=1}^{M} |v_j|^p \right)^{1/p} \, .$$

(3) A *p-norm* of a matrix A is defined as

$$\|A\|_p = \max_{x \neq 0} \frac{\|Av\|_p}{\|v\|_p} \, .$$

(4) Let A and B be square matrices of the same order. All matrix norms must have the properties

$$\|A\| \quad \geq 0, \quad \|A\| = 0 \text{ implies } A = 0$$
$$\|c \cdot A\| \quad = |c| \cdot \|A\|$$
$$\|A + B\| \leq \|A\| + \|B\|$$
$$\|A \cdot B\| \quad \leq \|A\| \cdot \|B\| \, .$$

(5) Special p-norms are

$$||A||_1 \ = \ \max_{j=1,\cdots,M} \sum_{i=1}^{M} |a_{ij}| \qquad \text{maximum column sum}$$

$$||A||_2 \ = \ \sqrt{\sigma(\overline{A}^{\mathrm{T}} \cdot A)}$$

$$||A||_\infty = \max_{i=1,2,\cdots,M} \sum_{j=1}^{M} |a_{ij}| \qquad \text{maximum row sum} \ ,$$

where $||A||_p$ is referred to as the L_p norm of A.

(6) In general $\sigma(A)$ does not satisfy the conditions in (4), so in general $\sigma(A)$ is *not* a true norm.

(7) When A is normal, $\sigma(A)$ *is* a true norm; in fact, in this case it is the L_2 norm.

(8) The spectral radius of A, $\sigma(A)$, is the lower bound of all the norms of A.

B. Some Properties of Tridiagonal Matrices

B.1 Standard Eigensystem for Simple Tridiagonal Matrices

In this work tridiagonal banded matrices are prevalent. It is useful to list some of their properties. Many of these can be derived by solving the simple linear difference equations that arise in deriving recursion relations.

Let us consider a *simple* tridiagonal matrix, i.e. a tridiagonal matrix with constant scalar elements a, b, and c; see Section 3.3.2. If we examine the conditions under which the determinant of this matrix is zero, we find (by a recursion exercise)

$$\det[B(M : a, b, c)] = 0$$

if

$$b + 2\sqrt{ac}\cos\left(\frac{m\pi}{M+1}\right) = 0 \quad , \quad m = 1, 2, \ldots, M .$$

From this it follows at once that the eigenvalues of $B(a, b, c)$ are

$$\lambda_m = b + 2\sqrt{ac}\cos\left(\frac{m\pi}{M+1}\right) \quad , \quad m = 1, 2, \ldots, M . \tag{B.1}$$

The right-hand eigenvector of $B(a, b, c)$ that is associated with the eigenvalue λ_m satisfies the equation

$$B(a, b, c)\vec{x}_m = \lambda_m \vec{x}_m \tag{B.2}$$

and is given by

$$\vec{x}_m = (x_j)_m = \left(\frac{a}{c}\right)^{(j-1)/2} \sin\left[j\left(\frac{m\pi}{M+1}\right)\right] \quad , \quad m = 1, 2, \ldots, M . \tag{B.3}$$

These vectors are the columns of the right-hand eigenvector matrix, the elements of which are

$$X = (x_{jm}) = \left(\frac{a}{c}\right)^{(j-1)/2} \sin\left[\frac{jm\pi}{M+1}\right] \quad , \quad \begin{array}{l} j = 1, 2, \ldots, M \\ m = 1, 2, \ldots, M \end{array} . \tag{B.4}$$

Notice that if $a = -1$ and $c = 1$,

$$\left(\frac{a}{c}\right)^{(j-1)/2} = e^{i(j-1)\pi/2} \ . \tag{B.5}$$

The left-hand eigenvector matrix of $B(a,b,c)$ can be written

$$X^{-1} = \frac{2}{M+1}\left(\frac{c}{a}\right)^{(m-1)/2} \sin\left[\frac{mj\pi}{M+1}\right] \ , \quad \begin{array}{l} m = 1,2,\dots,M \\ j = 1,2,\dots,M \ . \end{array}$$

In this case notice that if $a = -1$ and $c = 1$

$$\left(\frac{c}{a}\right)^{(m-1)/2} = e^{-i(m-1)\pi/2} \ . \tag{B.6}$$

B.2 Generalized Eigensystem
for Simple Tridiagonal Matrices

This system is defined as follows

$$\begin{bmatrix} b & c & & & \\ a & b & c & & \\ & a & b & & \\ & & & \ddots & c \\ & & & a & b \end{bmatrix} \begin{bmatrix} x_1 \\ x_2 \\ x_3 \\ \vdots \\ x_M \end{bmatrix} = \lambda \begin{bmatrix} e & f & & & \\ d & e & f & & \\ & d & e & & \\ & & & \ddots & f \\ & & & d & e \end{bmatrix} \begin{bmatrix} x_1 \\ x_2 \\ x_3 \\ \vdots \\ x_M \end{bmatrix} \ .$$

In this case one can show after some algebra that

$$\det[B(a - \lambda d, b - \lambda e, c - \lambda f] = 0 \tag{B.7}$$

if

$$b - \lambda_m e + 2\sqrt{(a - \lambda_m d)(c - \lambda_m f)}\cos\left(\frac{m\pi}{M+1}\right) = 0 \ , \quad m = 1,2,\dots,M \ .$$

If we define

$$\theta_m = \frac{m\pi}{M+1}, \quad p_m = \cos\theta_m$$

then

$$\lambda_m = \frac{eb - 2(cd + af)p_m^2 + 2p_m\sqrt{(ec - fb)(ea - bd) + [(cd - af)p_m]^2}}{e^2 - 4fdp_m^2} \ .$$

The right-hand eigenvectors are

$$\vec{x}_m = \left[\frac{a - \lambda_m d}{c - \lambda_m f}\right]^{(j-1/2)} \sin[j\theta_m] \ , \quad \begin{array}{l} m = 1,2,\dots,M \\ j = 1,2,\dots,M \ . \end{array}$$

These relations are useful in studying relaxation methods.

B.3 The Inverse of a Simple Tridiagonal Matrix

The inverse of $B(a, b, c)$ can also be written in analytic form. Let D_M represent the determinant of $B(M : a, b, c)$

$$D_M \equiv \det[B(M : a, b, c)] .$$

Defining D_0 to be 1, it is simple to derive the first few determinants, thus

$$\begin{align}
D_0 &= 1 \\
D_1 &= b \\
D_2 &= b^2 - ac \\
D_3 &= b^3 - 2abc .
\end{align}$$
(B.8)

One can also find the recursion relation

$$D_M = bD_{M-1} - acD_{M-2}$$
(B.9)

Equation (B.9) is a linear OΔE, the solution of which was discussed in Section 6.3. Its characteristic polynomial is $P(E) = E^2 - bE + ac$, and the two roots to $P(\sigma) = 0$ result in the solution

$$D_M = \frac{1}{\sqrt{b^2 - 4ac}} \left\{ \left[\frac{b + \sqrt{b^2 - 4ac}}{2} \right]^{M+1} - \left[\frac{b - \sqrt{b^2 - 4ac}}{2} \right]^{M+1} \right\}$$
$$M = 0, 1, 2, \dots ,$$
(B.10)

where we have made use of the initial conditions $D_0 = 1$ and $D_1 = b$. In the limiting case when $b^2 - 4ac = 0$, one can show that

$$D_M = (M + 1) \left(\frac{b}{2} \right)^M .$$

Then for $M = 4$

$$B^{-1} = \frac{1}{D_4} \begin{bmatrix}
D_3 & -cD_2 & c^2 D_1 & -c^3 D_0 \\
-aD_2 & D_1 D_2 & -cD_1 D_1 & c^2 D_1 \\
a^2 D_1 & -aD_1 D_1 & D_2 D_1 & -cD_2 \\
-a^3 D_0 & a^2 D_1 & -aD_2 & D_3
\end{bmatrix}$$

and for $M = 5$

$$B^{-1} = \frac{1}{D_5} \begin{bmatrix}
D_4 & -cD_3 & c^2 D_2 & -c^3 D_1 & c^4 D_0 \\
-aD_3 & D_1 D_3 & -cD_1 D_2 & c^2 D_1 D_1 & -c^3 D_1 \\
a^2 D_2 & -aD_1 D_2 & D_2 D_2 & -cD_2 D_1 & c^2 D_2 \\
-a^3 D_1 & a^2 D_1 D_1 & -aD_2 D_1 & D_3 D_1 & -cD_3 \\
a^4 D_0 & -a^3 D_1 & a^2 D_2 & -aD_3 & D_4
\end{bmatrix} .$$

The general element d_{mn} is

Upper triangle:
$$m = 1, 2, \ldots, M - 1 \quad, \quad n = m + 1, m + 2, \ldots, M$$
$$d_{mn} = D_{m-1} D_{M-n}(-c)^{n-m}/D_M$$

Diagonal:
$$n = m = 1, 2, \ldots, M$$
$$d_{mm} = D_{M-1} D_{M-m}/D_M$$

Lower triangle:
$$m = n + 1, n + 2, \ldots, M \quad, \quad n = 1, 2, \ldots, M - 1$$
$$d_{mn} = D_{M-m} D_{n-1}(-a)^{m-n}/D_M$$

B.4 Eigensystems of Circulant Matrices

B.4.1 Standard Tridiagonal Matrices

Consider the circulant (see Section 3.3.4) tridiagonal matrix

$$B_p(M : a, b, c) . \tag{B.11}$$

The eigenvalues are

$$\lambda_m = b + (a + c) \cos\left(\frac{2\pi m}{M}\right) - i(a - c) \sin\left(\frac{2\pi m}{M}\right) \quad,$$
$$m = 0, 1, 2, \ldots, M - 1 . \tag{B.12}$$

The right-hand eigenvector that satisfies $B_p(a, b, c)\vec{x}_m = \lambda_m \vec{x}_m$ is

$$\vec{x}_m = (x_j)_m = e^{ij(2\pi m/M)} \quad, \quad j = 0, 1, \ldots, M - 1 , \tag{B.13}$$

where $i \equiv \sqrt{-1}$, and the right-hand eigenvector matrix has the form

$$X = (x_{jm}) = e^{ij(2\pi m/M)} \quad, \quad \begin{array}{l} j = 0, 1, \ldots, M - 1 \\ m = 0, 1, \ldots, M - 1 \end{array} .$$

The left-hand eigenvector matrix with elements x' is

$$X^{-1} = (x'_{mj}) = \frac{1}{M} e^{-im(2\pi j/M)} \quad, \quad \begin{array}{l} m = 0, 1, \ldots, M - 1 \\ j = 0, 1, \ldots, M - 1 \end{array} .$$

Note that both X and X^{-1} are symmetric and that $X^{-1} = \frac{1}{M} X^H$, where X^H is the conjugate transpose of X.

B.4.2 General Circulant Systems

Notice the remarkable fact that the elements of the eigenvector matrices X and X^{-1} for the tridiagonal circulant matrix given by (B.11) do not depend on the elements a, b, c in the matrix. In fact, *all circulant matrices of order M have the same set of linearly independent eigenvectors*, even if they are completely dense. An example of a dense circulant matrix of order $M = 4$ is

$$
\begin{bmatrix}
b_0 & b_1 & b_2 & b_3 \\
b_3 & b_0 & b_1 & b_2 \\
b_2 & b_3 & b_0 & b_1 \\
b_1 & b_2 & b_3 & b_0
\end{bmatrix} . \tag{B.14}
$$

The eigenvectors are always given by (B.13), and further examination shows that the elements in these eigenvectors correspond to the elements in a complex harmonic analysis or complex discrete Fourier series.

Although the eigenvectors of a circulant matrix are independent of its elements, the eigenvalues are not. For the element indexing shown in (B.14) they have the general form

$$
\lambda_m = \sum_{j=0}^{M-1} b_j e^{i(2\pi j m/M)} ,
$$

of which (B.12) is a special case.

B.5 Special Cases Found from Symmetries

Consider a mesh with an even number of interior points. One can seek from the tridiagonal matrix $B(2M : a, b, a,)$ the eigenvector subset that has even symmetry when spanning the interval $0 < x \leq \pi$. For example, we seek the set of eigenvectors $\vec{x}_m$ for which

$$
\begin{bmatrix}
b & a & & & & & \\
a & b & a & & & & \\
 & a & \ddots & & & & \\
 & & & \ddots & a & & \\
 & & & a & b & a \\
 & & & & a & b
\end{bmatrix}
\begin{bmatrix}
x_1 \\ x_2 \\ \vdots \\ \vdots \\ x_2 \\ x_1
\end{bmatrix}
= \lambda_m
\begin{bmatrix}
x_1 \\ x_2 \\ \vdots \\ \vdots \\ x_2 \\ x_1
\end{bmatrix} .
$$

This leads to the subsystem of order M which has the form

$$B(M : a, \vec{b}, a)\vec{x}_m = \begin{bmatrix} b & a & & & & \\ a & b & a & & & \\ & a & \ddots & & & \\ & & \ddots & a & & \\ & & & a & b & a \\ & & & & a & b+a \end{bmatrix} \vec{x}_m = \lambda_m \vec{x}_m \ . \qquad \text{(B.15)}$$

By folding the known eigenvectors of $B(2M : a, b, a)$ about the center, one can show from previous results that the eigenvalues of (B.15) are

$$\lambda_m = b + 2a \cos\left(\frac{(2m-1)\pi}{2M+1}\right) \ , \quad m = 1, 2, \ldots, M \qquad \text{(B.16)}$$

and the corresponding eigenvectors are

$$\vec{x}_m = \sin\left(\frac{j(2m-1)\pi}{2M+1}\right) \ , \quad j = 1, 2, \ldots, M \ .$$

Imposing symmetry about the same interval but for a mesh with an odd number of points leads to the matrix

$$B(M : \vec{a}, b, a) = \begin{bmatrix} b & a & & & & \\ a & b & a & & & \\ & a & \ddots & & & \\ & & & \ddots & a & \\ & & & a & b & a \\ & & & & 2a & b \end{bmatrix} \ .$$

By folding the known eigenvalues of $B(2M - 1 : a, b, a)$ about the center, one can show from previous results that the eigenvalues of (B.16) are

$$\lambda_m = b + 2a \cos\left(\frac{(2m-1)\pi}{2M}\right) \ , \quad m = 1, 2, \ldots, M$$

and the corresponding eigenvectors are

$$\vec{x}_m = \sin\left(\frac{j(2m-1)\pi}{2M}\right) \ , \quad j = 1, 2, \ldots, M \ .$$

B.6 Special Cases Involving Boundary Conditions

We consider two special cases for the matrix operator representing the 3-point central difference approximation for the second derivative $\partial^2/\partial x^2$ at all points away from the boundaries, combined with special conditions imposed at the boundaries.

Note: In both cases

$$m = 1, 2, \ldots, M$$
$$j = 1, 2, \ldots, M$$
$$-2 + 2\cos(\alpha) = -4\sin^2(\alpha/2) \ .$$

When the boundary conditions are Dirichlet on both sides,

$$
\begin{bmatrix}
-2 & 1 & & & \\
1 & -2 & 1 & & \\
& 1 & -2 & 1 & \\
& & 1 & -2 & 1 \\
& & & 1 & -2
\end{bmatrix}
\qquad
\begin{aligned}
\lambda_m &= -2 + 2\cos\left(\frac{m\pi}{M+1}\right) \\
\vec{x}_m &= \sin\left[j\left(\frac{m\pi}{M+1}\right)\right] \ .
\end{aligned}
\qquad \text{(B.17)}
$$

When one boundary condition is Dirichlet and the other is Neumann (and a diagonal preconditioner is applied to scale the last equation),

$$
\begin{bmatrix}
-2 & 1 & & & \\
1 & -2 & 1 & & \\
& 1 & -2 & 1 & \\
& & 1 & -2 & 1 \\
& & & 1 & -1
\end{bmatrix}
\qquad
\begin{aligned}
\lambda_m &= -2 + 2\cos\left[\frac{(2m-1)\pi}{2M+1}\right] \\
\vec{x}_m &= \sin\left[j\left(\frac{(2m-1)\pi}{2M+1}\right)\right] \ .
\end{aligned}
\qquad \text{(B.18)}
$$

C. The Homogeneous Property of the Euler Equations

The Euler equations have a special property that is sometimes useful in constructing numerical methods. In order to examine this property, let us first inspect Euler's theorem on homogeneous functions. Consider first the scalar case. If $F(u, v)$ satisfies the identity

$$F(\alpha u, \alpha v) = \alpha^n F(u, v) \tag{C.1}$$

for a fixed n, F is called homogeneous of degree n. Differentiating both sides with respect to α and setting $\alpha = 1$ (since the identity holds for all α), we find

$$u\frac{\partial F}{\partial u} + v\frac{\partial F}{\partial v} = nF(u, v) . \tag{C.2}$$

Consider next the theorem as it applies to systems of equations. If the vector $F(Q)$ satisfies the identity

$$F(\alpha Q) = \alpha^n F(Q) \tag{C.3}$$

for a fixed n, F is said to be homogeneous of degree n and we find

$$\left[\frac{\partial F}{\partial q}\right] Q = nF(Q) . \tag{C.4}$$

Now it is easy to show, by direct use of (C.3), that both E and F in (2.11) and (2.12) are homogeneous of degree 1, and their Jacobians, A and B, are homogeneous of degree 0 (actually the latter is a direct consequence of the former).[1] This being the case, we notice that the expansion of the flux vector in the vicinity of t_n, which, according to (6.105), can be written in general as

$$E = E_n + A_n(Q - Q_n) + O(h^2)$$
$$F = F_n + B_n(Q - Q_n) + O(h^2) , \tag{C.5}$$

becomes

---

[1] Note that this depends on the form of the equation of state. The Euler equations are homogeneous if the equation of state can be written in the form $p = \rho f(\epsilon)$, where ϵ is the internal energy per unit mass.

$$E = A_n Q + O(h^2)$$
$$F = B_n Q + O(h^2) \, , \tag{C.6}$$

since the terms $E_n - A_n Q_n$ and $F_n - B_n Q_n$ are identically zero for homogeneous vectors of degree 1 (see (C.4)). Notice also that, under this condition, the constant term drops out of (6.106).

As a final remark, we notice from the chain rule that for *any* vectors F and Q

$$\frac{\partial F(Q)}{\partial x} = \left[\frac{\partial F}{\partial Q} \right] \frac{\partial Q}{\partial x} = A \frac{\partial Q}{\partial x} \, . \tag{C.7}$$

We notice also that for a homogeneous F of degree 1, $F = AQ$ and

$$\frac{\partial F}{\partial x} = A \frac{\partial Q}{\partial x} + \left[\frac{\partial A}{\partial x} \right] Q \, . \tag{C.8}$$

Therefore, if F is homogeneous of degree 1,

$$\left[\frac{\partial A}{\partial x} \right] Q = 0 \tag{C.9}$$

in spite of the fact that individually $[\partial A / \partial x]$ and Q are not equal to zero.

Index